Exploring

CALCULUS

WITH MAPLE®

Mark H. Holmes
Joseph G. Ecker
William E. Boyce
William L. Siegmann

Rennsselaer Polytechnic Institute

Addison-Wesley Publishing Company
Reading, Massachusetts • Menlo Park, California • New York
Don Mills, Ontario • Wokingham, England • Amsterdam • Bonn
Sydney • Singapore • Tokyo • Madrid • San Juan • Milan • Paris

ISBN: 0-201-52616-6

3 4 5 6 7 8 9 10-MU-959493

Preface

This book contains a sequence of Laboratories that can be used to integrate Maple®[†] into a calculus course. We assume minimal experience with Maple and have therefore tried to make the book self-contained. So, a chapter is included that summarizes the Maple commands that are used in the book. We do assume, however, that the student can start Maple, knows how to enter commands, and can print plots and text files. Information about the hardware configuration, the operating system, plotting and printing must be locally supplied.

The Laboratories are designed to be used in conjunction with a textbook, but are independent of the particular text that is selected. The Lab topics should fit into almost any calculus course, although perhaps in a somewhat different order than they appear in this book. This should not cause difficulty inasmuch as the Labs are essentially independent of one another. There are more than enough Laboratories for a two- or three-semester course, thus affording some choice to an instructor. Many Labs also contain several problems, not all of which need be assigned.

The Laboratories have been written so the first few (1–8) offer more guidance on using Maple than do those that follow. However, as new commands are introduced in the later Labs, they are explained in the examples preceding the lab problems. For those who may bounce around in the book, some of the explanations of the commands are repeated, or else there is a reference to the Lab where an

---

[†] Maple® is a registered trademark of Waterloo Maple Software.

example using the command is given. From time to time certain Labs fit together nicely, and in this case this is pointed out in the Lab statement.

Ideally we try to have each Lab assignment take between two and three hours. However, as you will see, the Labs vary significantly in their length. In some Labs we have provided more than three hours worth, to give instructors options on what to select. We have also tried to divide the longer ones into independent parts, so if time is an issue an instructor can omit certain problems.

It should also be mentioned that at Rensselaer we do not limit the use of Maple to the Laboratories. We use it in lectures to help with problem solving and, more importantly, with visualizing the behavior of functions. Students also have Maple available for use in regular homework assignments if they wish.

Ideas for a few of the Laboratories were developed while visiting some familiar places. These places are mentioned in the labs, and we would like to thank the McDonald's Corporation and the Imax Corporation for giving us permission to use their names in the Labs.

The Maple calculus project began at Rensselaer in 1988. For the first couple of years it consisted primarily of pilot sections, but starting in the Fall semester of 1991 all of the (approximately 1000) students in our calculus courses were involved. Many of the Labs in this book have been used in this course, although they have been spruced up somewhat for publication. Also, most were written exclusively by us, although we had some particularly helpful input, or ideas, from several individuals. These include Bernard Fleishman, Edith Luchins, and Joseph Pimbley. We would like to thank them for their contributions. One person to whom we owe particular thanks is Edward Damiano, who read the entire manuscript and made many valuable contributions to the book. We would also like to thank James D. Meindl, the Provost and Senior Vice President for Academic Affairs at Rensselaer, for his continued and vigorous support of the computer calculus project from its inception.

Mark H. Holmes
Joseph G. Ecker
William E. Boyce
William L. Siegmann

Troy, New York
August, 1992

Table of Contents

Part 3   Sample Laboratory   27

Part 4   Laboratories   35

Exploring

CALCULUS

WITH MAPLE®

Part 1

Introduction to Using Maple in Calculus

First, we would like to say welcome! This small book contains some of our ideas on how to use Maple to discover the richness and power of calculus. We hope you enjoy it.

It is surprising what little impact computers have had on calculus. It is surprising not only because of the profound impact computers have had in our everyday lives but also because they have the potential to relieve us of having to carry out very tedious, and boring, manipulations. By eliminating the drudgery of routine manipulations there is more time to concentrate on the more interesting aspects of the subject. This includes problem formulation, the testing of hypotheses, and the analysis of the results. Moreover, the graphical and computing capabilities of Maple can be used to develop an understanding of the fundamental concepts of calculus. Another benefit is that a computer algebra system like Maple can help in getting you away from the traditional passive classroom environment and allows you to discover the beauty and strength of calculus more actively.

Our primary objective with this book is to integrate symbolic computing and computer graphics into calculus to facilitate the understanding of concepts and to expand the range and variety of applications. In conjunction with this we intend to equip you with powerful problem solving tools and to promote confidence and good judgment in their use. We also hope to reduce tedious manipulations so more time can be spent on problem formulation and the analysis of the results.

1

We also have non-goals. Namely, we do not intend to train you to be a computer programmer, to teach you the subtleties of a computer operating system (such as UNIX), or to use computers to solve routine homework problems. At the same time, you must be aware that a certain mastery of the computer and Maple is necessary. However, we have written the Laboratories so that you will learn the commands of Maple in a gradual and timely manner. In this way you are able to concentrate on the mathematical problems, and not the maze of problems that can arise when using a computer system.

It is important to stress that, despite its power, the effective use of Maple requires clear thinking and good judgment. The fact is that Maple is a tool and it is up to you to employ it wisely. The benefits of this approach are enormous as Maple can help develop imaginative and critical thinking, rather than having you concentrate on routine mathematical operations.

1. Effects of Maple on a Calculus Course

The impact of Maple on a calculus course depends on the level of participation and the emphasis of the course. The following outline will give you some idea of the changes we have seen in our program since the introduction of Maple:

1. Course becomes *more quantitative* as some tasks become routine, e.g.,
   a. solving equations,
   b. evaluation of integrals,
   and as other tasks become feasible, e.g.,
   c. evaluation of Riemann sums,
   d. evaluation of partial sums of infinite series.
   Because of this, the use of calculus for constructing approximations becomes more important.

2. Course becomes *more visual* as graphs become easier to draw.
   a. Graphs can sometimes replace, or suggest, analysis. E.g., they can be used to:
      (i)   show the existence, and approximate location, of max/min points;
      (ii) establish monotonicity and generate graphs of inverse functions;
      (iii) and study the errors in Taylor approximations.
   b. The idea of proper scaling of graphs becomes essential, so that important features are made apparent.
   c. Sequences of numbers or functions can now be plotted to aid in understanding convergence, e.g., partial sums of an infinite series.
   d. Curves and surfaces can be plotted in three dimensions.

3. Certain topics get more "air time." Examples include:
   a. Parametric representations of curves and surfaces,

e.g., in sketching conic sections or quadric surfaces.
b. Use of Taylor series for applications,
   e.g., in differential equations, and function approximation.
   While others get less. E.g.,
c. integration techniques and the detailed properties of conics.

4. More emphasis can be given to
   a. formulation (modeling) of problems,
   b. problem solving strategies,
   c. interpretation of solutions.
   For instance, one can investigate
   d. the dependence of solutions on one or more parameters,
   e. the importance of limiting cases,
   f. variations of a single problem, as additional features are included in the
      model, e.g., the trajectory of baseball with and without air resistance.
   Finally, examples and problems can be chosen because they are interesting,
   rather than because the solution works out simply.

5. Some time must be spent on Maple instruction, e.g.,
   a. the syntax of commands,
   b. the saving, editing, and printing files (including both text and plots).

6. Miscellaneous:
   a. Questions related to changes in a problem can often be answered
      quickly.
   b. Examples and problems, such as those relating to topics such as arc
      length, surface area, and curvature, do not have to be chosen quite so
      carefully.
   c. The integral test for convergence of a series becomes more important
      because of Maple's powerful integrator.
   d. Lab assignments can have the somewhat unexpected benefit of
      improving the student's technical writing abilities.

---

2. Maple Calculus at Rensselaer

We by no means expect everyone to follow our path into the "Maplefication" of
calculus, although we highly recommend it. However, so you will have an idea of
what we have done and how we use Maple, in terms of Laboratory assignments and
classroom instruction, we will describe our course as it now stands. We have three
different entry-level calculus courses, each designed for students with different
backgrounds. One of these courses assumes a basic knowledge of limits and
derivatives, so it starts further along in the traditional calculus sequence.
Consequently, you will find Labs in this book that are introductory, yet cover

material a few chapters into most calculus textbooks (by introductory we mean that the examples are more explanatory in terms of how to use Maple). We have left them as introductory because others may also have such courses, others may not start using Maple until later in the term, and some students may need a little more help getting started with Maple. We find that it usually takes two or three Labs before students are comfortable with using Maple, the computer, and constructing Laboratory write-ups.

Each calculus course meets in a computer lab for one hour each week. We normally hand out the Lab assignment a day or two before the Lab session meets, and we give the students a week to work on it (faculty vary somewhat in exactly how they do this). Our guidelines for the lab write-ups are given in the next section. We stress upon the student the importance of handing in a readable and clearly presented write-up. It is often this aspect of the course, more than using Maple, that causes them the most trouble in the beginning. However, as the course progresses they become very good at writing up the Labs.

The computer labs are run by TAs, and there is usually an undergraduate assistant present to help. The TA spends 20 minutes or so going over the examples in the Lab and commenting on some of the more subtle aspects of the assignment. The rest of the time the students work on the Lab, with the TA and undergraduate assistant walking around and helping as necessary. We allow them to work in teams as long as they clearly identify the team members (our guidelines for collaboration are given in the next section). Our TAs are also expected to hold one of their office hours in the lab and these hours are made available to everyone taking calculus. The one lab session that is different from the rest is the first one. This is when they are introduced to Maple and the computer. They are shown how to print, where to find help, how to stop, quit, and then restart a Maple session, etc.

3. Guidelines for Laboratory Reports and Collaboration

The following are only meant to be suggestive and your instructor will indicate exactly what is expected for each Lab assignment.

GUIDELINES FOR MAPLE LABORATORY REPORTS

Maple Laboratory reports normally require both Maple computer output and a few comments or explanations.

Computer output should be a listing of one or more terminal sessions, possibly including computer drawn plots. The listing(s) should:

- Contain all the Maple commands you used, along with the Maple responses, to answer the questions.

- Identify clearly all plots. This includes labeling the axes and identifying individual curves (in some cases this will have to be done by hand).

- Be edited to indicate what you want graded. Minimally, you should cross out with a pen or pencil those portions of the session you want to be omitted. In general, your paper must be easily readable.

Written output should contain your answers to questions where explanations, comparisons, tables, conclusions, etc., are requested. These should be written clearly in English sentences on a separate sheet or in the margin of your printout, as indicated by your instructor. You should *remember* that neatness counts!

GUIDELINES FOR COLLABORATION

1. You may prepare the Maple session for the Lab by yourself or jointly with *one* other student. Your write-up for the Lab may also be submitted either individually or jointly.

2. Only two students may submit a joint Lab session or write-up. It is not acceptable for more than two students to hand in the same Lab report.

3. You must indicate on your Lab report whether you are handing in joint work, and if so, who is your partner.

4. You are free to consult with your instructor, the course assistants, consultants, and other students about Maple commands needed for the project. However, conclusions and explanations must be your team's own ideas, and the write-ups and computer sessions must be your team's own work.

---

4. Hardware Requirements

It goes without saying that each student needs to have access to a computer and the computer has to have access to Maple. The harder question to answer is what type of computer should be used. Because of the graphical nature of many of the Labs, it is strongly recommended, if not essential, that the computer have a graphical display capability (color is not necessary). All the Labs in this book have been run with no difficulty on a Macintosh IIci running system 7.0.1 with 8 MB of RAM. To run the Labs we set the memory allocation for Maple at 5000K, although no Lab took more than 2000K. We should also point out that the Labs were originally developed on a SUN 4/40 and an IBM RS6000, that were running X-windows.

When developing the Labs we used Maple V, Release 1. Although earlier versions of Maple may work, we recommend using Maple V. Inevitably there will be new releases of Maple while this book is in print, and new features may make some of the Labs easier or open up new avenues of investigation. Of course, we recommend taking advantage of this.

Part 2

Introduction to Maple

This chapter introduces you to Maple and how to use it to solve mathematical problems. The information in this chapter is not intended to be consumed in one sitting. Rather it should serve as a reference as you work through the Laboratories. It contains most of what you need to know to do all of the projects in this book, but, at the same time, it is impossible in a short chapter to detail all the facets of a symbolic computing language that is as sophisticated as Maple. One very valuable source of information is Maple's on-line help facility, and this is discussed later in the chapter. You should also be aware that there are several manuals and help guides for Maple. These are not needed for the Laboratories, but a few are listed at the end of the chapter for your information.

To help you get started, an introductory demonstration using Maple is given in the next section. Its primary goal is to show how Maple works and to illustrate some of its basic capabilities. This simple introduction is sufficient to get you started, as we have written the Laboratories so you are introduced to new commands as you need them. Also, the first few Labs are designed to help you become familiar with Maple and to guide you through this development stage. What this all means is that you do not need to memorize a bunch of random Maple commands or wade through pages and pages of a reference manual before starting the Labs.

There are, however, a few things you should think about before starting. One is that you should figure out how to print a Maple session and the plots you create (this is something that will depend on your computer system). Another useful skill is the ability to edit a Maple command. We all make mistakes, and in Maple you can go back, edit a command, and then reissue it. This is discussed in Section 4 of this chapter. A third skill worth developing is how to stop a Maple session, go eat lunch or maybe even go to class, and then return and restart the session as though you had never left. This is explained in Section 3.

There are two other sections in this chapter worth mentioning. In Section 5 there is a discussion of certain commands in Maple that can cause confusion for someone just beginning. The section after that is about troubleshooting. One of the fundamental laws of computing states that occasionally you will enter a perfectly good command and either get absolutely no response or else some nonsense that looks as though it belongs in someone else's Maple session. Well, it is impossible to cover all possibilities for this situation, but a few helpful suggestions to get you out of this predicament are given in Section 6.

---

1. Introductory Demonstration of Maple

The best way to learn how to use a computer program is to use it to solve a problem or two. Maple is no different, and we will use this approach to demonstrate its basic capabilities. It is very important that you are not a passive participant when you read this section. You should start up Maple on your computer and work through the steps with the demonstration (however, it isn't necessary to include the comment lines). After this it would then be worthwhile to experiment with Maple just to see what you can, and cannot, do.

PROBLEM

Solve the equation $2x^2 - 13x - 24 = 0$.

The following is a Maple session that solves this problem. Note that each Maple command ends with a semicolon (;) and the computer does not execute the command until the line is entered (by hitting the return key or, on some machines, the enter key). Also, any line beginning with a # symbol indicates to Maple that this is a comment line, and it is therefore ignored. Finally, the symbol > beginning each line is the Maple prompt (on some machines the prompt symbol is a bullet •).

```
        | \ ^ / |        MAPLE V
 . _ | \ |     | / | _ .  Copyright (c) 1981-1990 by the University of Waterloo.
   \    MAPLE    /        All rights reserved.  MAPLE is a registered trademark of
   <_____ _____>       Waterloo Maple Software.
          |              Type ? for help.
>#  We first enter the function  y = 2*x^2 – 13*x – 24  (note the way powers
>#  and multiplication are expressed here).
```

> y := 2*x^2 − 13*x − 24;

$$y := 2\,x^2 - 13\,x - 24$$

># Maple has echoed our command (in this book all Maple responses will appear
># in italics). Also, note that in Maple one uses := to indicate assignment. In this
># example the name y has been assigned to the polynomial 2*x^2 − 13*x − 24 .
># Hereafter we can refer to the polynomial simply as y .

># Let's experiment with y a little before solving the problem. First we
># can try evaluating y at x = 0 . This can be accomplished as follows:
> x := 0;

$$x := 0$$

> y;

$$-24$$

># We get the expected result. However, we are now stuck with x set to
># zero. To return it to its old unevaluated self, use the following command:
> x := 'x';

$$x := x$$

> y;

$$2\,x^2 - 13\,x - 24$$

># There is another way to evaluate y, which doesn't require us to unevaluate x .
># This is the substitution command (`subs`), and it is
> subs(x = 0, y);

$$-24$$

># Now let's plot y from x = −3 to x = 10 . This is done using the
># `plot` command as follows:
> plot(y, x = −3..10, title = `Plot of  y = 2*x^2 − 13*x − 24`);

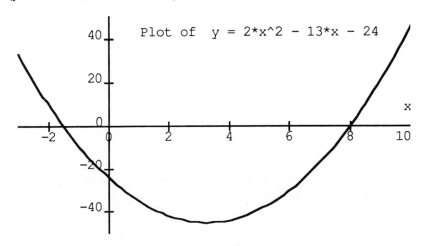

># Perhaps the most unusual aspect of the plot command is the way the x
># interval is entered. Also, note we have used back quotes (`) around
># the title of the plot. If you ever forget the specifics of a command you can
># always use the on-line help, which is accessed using the ? symbol. To
># find out about `plot` we would use the following (a semicolon isn't
># necessary when using ?):
> ?plot

># The Maple response hasn't been given here, but you can find out what it
># is easily enough by simply entering this command.

># We now return to the original problem of solving the equation. In the
># plot we can see that one root is between −2 and 0, while the other
># is very near 8 . To find out exactly what they are we use the
># `solve` command:
> solve(y = 0, x);

$$8, \ -3/2$$

># So Maple has given us answers we expected based on the plot.
># Note that in this command we have specified the equation (y = 0) and
># the variable to solve for (x).

># We now end this Maple session using the command:
> quit

---

2. Maple Commands

The following pages contain a succinct summary of many of the Maple commands
that are used in this book. It is worth looking them over briefly to get an overview
of what can be done. However, it is not worth the time to try to memorize the lot
when you are first starting out. You will learn them as you work through the
Laboratories. It should also be pointed out that this listing is not intended to replace
the manual but only to serve as a guideline, and perhaps a quick reference, for what
needs to be used in the Laboratories. For more extensive assistance it is
recommended that you use the on-line help (see the ? command described in the
listing) or else consult one of the Maple manuals listed at the end of this chapter.

As a reminder, a semicolon (;) or colon (:) must follow every command.
Also, anything on a line following the symbol # is ignored by Maple (this is the
way comments are added to Maple text).

In the following summary the command is given, then there is a short
description of the command, and after that there are usually a few examples.

OPERATORS (?+)

addition: +   subtraction: −

multiplication: *   division: /

exponentiation: ^ or **

repetition: $

Example: diff(x^5 , x$3) yields $60\,x^2$

CONSTANTS (?PI)

π : Pi ∞ : infinity $\sqrt{-1}$: I

ELEMENTARY FUNCTIONS (?EXP)

exponential: exp(x)   natural logarithm: ln(x)

absolute value: abs(x)   square root: sqrt(x) or x^(1/2)

trigonometric: sin(x) , cos(x) , tan(x) , sec(x) , cot(x) , csc(x)

The trigonometric functions in Maple require angles measured in radians.

COMMANDS USED IN LABORATORIES 1 – 25

> ?   Provides information and examples about Maple commands. The ? can be used for help for any Maple command and package.

?   # gives information about the help facility

?library   # gives the standard commands and functions

?index   # gives an index of help descriptions

?plot   # gives information on the `plot` command

?plot[polar]   # gives information on the plot option of using polar coordinates

> ";   This gives the previously computed result. Maple remembers the previous three (i.e., you can use " and "" and """).

```
f := exp(2);
g := 6*";                   # gives  6 exp(2)
h := "*"";                  # gives  6 exp(2)²
```

> x := 'x';   Returns x to a variable (note back quotes are **not** used here).

> f := x–>stuff;   Arrow notation to define f as a function of x . For help on this
                     use ?–> .

$$f := x \to 3*x + 5;$$

$$f(2);   \# \text{ gives } 11$$

$$f := 'f';   \# \text{ unassigns the previous definition for f}$$

$$g := (x,y) \to x*y^2;$$

> diff(f,x$n);   Finds the n-th derivative of f with respect to x.

$$\text{diff}(f, x);   \# \text{ this gives the first derivative } f'$$

$$\text{diff}(f, x\$3);   \# \text{ this gives the third derivative } f'''$$

$$\text{diff}(f, x, x, x); \# \text{ this also gives the third derivative } f'''$$

$$\text{diff}(x^3, x);   \# \text{ gives } 3\,x^2$$

$$\text{diff}(t^3, t\$3);   \# \text{ gives } 6$$

$$f := x \to 2*x^3 + 5;$$

$$\text{diff}(f(x), x);   \# \text{ gives } 6\,x^2$$

> Digits := n;   Sets the number of digits used for floating point numbers to n (the
                     default is 10). Note the capital D in this command.

> evalf(f);   Evaluates the expression f using floating point arithmetic.

$$\text{evalf}(-1/4 + \text{sqrt}(33)/4);   \# \text{ gives } 1.186140662$$

$$\text{evalf}(Pi, 20);   \# \text{ gives } 3.1415926535897932385$$

> expand(f);   Expands the expression f using the laws of algebra and
                     trigonometry.

$$\text{expand}(\ln((x + 2)/x^2)); \# \text{ gives } \ln(x + 2) - 2\ln(x)$$

$$\text{expand}((s + 1)*(s + 3)); \# \text{ gives } s^2 + 4\,s + 3$$

> factor(f);   Factors the given expression.

$$\text{factor}(x^2 + 5*x + 6);   \# \text{ gives } (x + 3)\,(x + 2)$$

> fsolve(f = a, x); Solves the equation f = a for x . The answer is given in decimal form. Usually `fsolve` returns a single real root, but for some polynomial and transcendental equations it will find all real roots.

fsolve(x^2 – 3*x + 2, x);   # gives 1.000000000, 2.000000000

fsolve(r^3 + 4 = 45, r);   # gives 3.448217240

fsolve(f = a, x, x1..x2);   # This solves f(x) = a for x1 < x < x2.

> int(f, x);   Finds the indefinite integral of f with respect to x . The arbitrary constant of integration is not included in the answer.

int(x^2, x);   # gives $1/3\ x^3$

> int(f, x = a..b); Finds the definite integral of f from a to b . When this command is followed by evalf(") the integral is evaluated numerically.

int(x^2, x = 0..2);   # gives 8/3

int(exp(–z^2), z = 0..1); # gives $1/2\ Pi^{1/2}\ erf(1)$

evalf(");   # gives .7468241330

> leftbox(f, x = a..b, n);   Plots rectangular boxes used to approximate the definite integral of f over $a \le x \le b$. Height of each box determined by the value of the function at the left side of each subinterval; n specifies the number of boxes.

with(student):

leftbox(sin(x),x=0..Pi/2,10);

> leftsum(f, x = a..b, n);   Finds the approximation of the definite integral of f over $a \le x \le b$ when leftboxes are used; n specifies the number of boxes (or subintervals).

with(student):

leftsum(sin(x), x=0..Pi/2, 10);

evalf(value("));   # gives .9194031700

leftsum(sin(x), x=0..Pi/2, 10);

evalf(value("));   # gives .9194031700

> limit(f, x = a);   Finds the limit of f as $x \rightarrow a$. On some expressions it is best to use the `expand` command first before using `limit.`

limit(sin(x)/x, x = 0);   # gives 1

limit(exp(b), b = infinity);   # gives infinity

limit(−1/x, x = 0, right);   # gives −infinity

> middlesum(f, x = a..b, n);   Finds the approximation of the definite integral of f over $a \le x \le b$ using middleboxes; n specifies the number of boxes (or subintervals).

with(student):

middlesum(sin(x), x = 0..Pi/2, 10);

evalf(value("));   # gives 1.001028825

> plot(f, x = x1..x2, title = `example`);   Plots the graph of f(x) for $x1 \le x \le x2$. Note the backquotes around the plot title.

plot(cos(x), x = 0..Pi);   # This plots $y = \cos(x)$
                                    # for $0 \le x \le \pi$.

plot(f,x=2..4,0..1, title =`y=f(x)`);   # This restricts $2 \le x \le 4$,
                                    # $0 \le y \le 1$.

> plot({f, g}, x = x1..x2, title =`TEST`);   This is used for plotting 2 (or more) graphs on the same axes.

plot({x^2, sin(x)}, x=0..Pi, −1..3);   # this plots y = x^2 and y =
                                    # sin(x) for $0 \le x \le \pi$,
                                    # $−1 \le y \le 3$.

> plot([f, g, t = a..b], title =`test`);   Plots the parametric equations x = f(t) , y = g(t) for $a \le t \le b$.

plot([sin(t), cos(t), t= −Pi..Pi], title = `A Circle`);

plot([f, g, t = 0..1], −2..3);   # This restricts $−2 \le x \le 3$.

plot([f, g, t = 0..1], –2..3, 0..5);   # Now –2 ≤ x ≤ 3, 0 ≤ y ≤ 5.
plot([r, t, t = 0..1], coords = polar); # This plots the polar equation
# r = r(t).
plot({[f, g, t=0..1], [F, G, t=–1..3]}, title=`A Couple of Curves`);

This plots two
parametric curves.

> quit   Quit Maple. If this doesn't work, try ;;;quit .

> rightbox(f, x = a..b, n);   Plots rectangular boxes used to approximate the definite
integral of f over a ≤ x ≤ b . Height of each box
determined by the value of the function at the right side
of each subinterval; n specifies the number of boxes.

with(student):

rightbox(sin(x), x = 0..Pi/2, 10);

> rightsum(f, x = a..b, n);   Finds the approximation of the definite integral of f
over a ≤ x ≤ b when rightboxes are used; n specifies
the number of boxes (or subintervals).

with(student):

rightsum(sin(x), x = 0..Pi/2, 10);

evalf(value("));   # gives 1.076482803

> simplify(f);   Simplifies the expression f . On some expressions it is best first
to use the `expand` command before using `simplify.`

simplify(16^(1/2) + 6);   # gives 10

simplify(exp(a) + ln(b*exp(c))); # gives exp(a) + ln(b) + c

> simpson(f, x = a..b, n);   Finds the approximation of the definite integral of f
over a ≤ x ≤ b using Simpson's rule; n specifies
number of subintervals (it must be even).

with(student):

simpson(sin(x), x = 0..Pi/2, 10);

evalf(value("));   # gives 1.000003392

> solve(f = a, x); This solves $f(x) = a$ for x . This command produces exact solutions, if available, while `fsolve` produces numerical answers.

solve(sin(x) + y = 2, x);   # gives $-\arcsin(y - 2)$

solve(x^2 + 2*x*y = 1, x);   # gives $-y + (y^2 + 1)^{1/2}$,

$-y - (y^2 + 1)^{1/2}$

sol := solve(x^2 – 9 = 0, x);

sol[1];   # gives 3

sol[2];   # gives –3

solve({x + y = 1, 2*x + y = 3}, {x,y});

gives $\{y = -1 , x = 2\}$

> subs(x = x0, f);   Substitutes x0 for x in the expression f . Note that x0 can be either a numerical value or an algebraic expression. It is not necessary to use x := 'x' after this command.

subs(x = y^3, x^2 + 9*x);   # gives $y^6 + 9\,y^3$

subs(x = 0, y = –1, z = Pi, x + y + cos(z));   # gives $-1 + \cos(Pi)$

> sum(f, i = m..n);   Calculates the sum of f from i = m to i = n (m , n may be negative).

sum(i^3, i =1..3);   # gives 36

f := 2*i + 1;

sum(f, i = –1..4);   # gives 24

> trapezoid(f, x = a..b, n);   Finds the approximation of the definite integral of f over $a \le x \le b$ using trapezoidal rule; n specifies the number of subintervals.

with(student):

trapezoid(sin(x), x = 0..Pi/2, 10);

evalf(value("));   # gives .9979429868

ADDITIONAL COMMANDS USED IN LABORATORIES 26–39

> angle(u, w);   Gives the angle between the vectors u and w .[†]

$$\text{angle(vector([1, 0, 0]), vector([1, 1, 1]));} \quad \# \ \arccos(1/3 \ 3^{1/2})$$

> convert(T, polynom);   Converts a Taylor series T to a polynomial.

$$s := \text{taylor}(\sin(x), x, 5); \quad \# \text{ gives } s := x - 1/6 \ x^3 + O(x^5)$$
$$p := \text{convert}(s, \text{polynom}); \ \# \text{ gives } p := x - 1/6 \ x^3$$

> convert(v, list);   Converts a vector v to a list that can then be used in
                              `plot3d.`

> convert(x, degrees);   Converts x from radians to degrees.

$$\text{convert(Pi, degrees);} \qquad\qquad \# \text{ gives } 180 \text{ degrees}$$

> crossprod(v, w);
                              Computes the cross-product of the vectors v and w .[†]

$$v1 := \text{vector([1, 2, 3]); } \ v2 := \text{vector([2, 3, 4]);}$$
$$\text{crossprod}(v1, v2); \qquad\qquad \# \text{ gives } [-1, 2, -1]$$

> display3d({F, G});   Command used to display multiple 3D objects.

with(plots):
C := plot3d([cos(t), sin(t), t], t = 0..Pi, s=0..1, grid = [35, 2]):
P := plot3d([x, y, x + Pi/2], x = −1..1,y = 0..2):
S := plot3d(sin(x + y), x = 0..2, y = −1 .. 1):
display3d({C, P, S}, title = `A Curve and a Couple of Surfaces`);

[†] The command `with(linalg):` must appear somewhere in Maple session before this command. Note: This applies wherever the † symbol appears in this section.

> dotprod(v, w); Calculates the dot product of the vectors v and w .[†]

    v:=vector([1, x, y]);   w:=vector([1, 0, 2]);

    dotprod(v, w);   # gives $1 + 2\,y$

    dotprod(vector([1, 2]),vector([a, b]));   # gives $a + 2\,b$

> evalm(a*v + b*w);   Calculates $av + bw$ where a, b are scalars and v, w are vectors.[†]

    v:=vector([1, x, y]);   w:=vector([1, 0, 2]);   u:=vector([−1, 0, q]);

    evalm(2*v − 3*w);   # gives $[−1, 2\,x, 2\,y − 6]$

    evalm(v + w − u / 2);   # gives $[5/2, x, y + 2 − 1/2\,q]$

> grad(f, [x, y, z]);   Finds the gradient of f .[†]

    grad(x*y*z, [x, y, z]);   # gives $[y\,z, x\,z, x\,y]$

    grad(r*t*sin(s), [r, s, t]);   # $[t\,\sin(s), r\,t\,\cos(s), r\,\sin(s)]$

> map(diff, v, x);   Differentiates the vector v with respect to x .[†]

    v := vector([3*x, cos(x^2)]);

    map(diff, v, x);   # gives $[3, −2\,x\,\sin(x^2)]$

> mtaylor(f, [x, y], n);   Computes the multivariate Taylor series of f to order n − 1, in the variables x and y .

    readlib(mtaylor):

    mtaylor(sin(x + y), [x, y], 2);   # gives $x + y$

    mtaylor(cos(x + y), [x, y], 3);   # gives $1 − 1/2\,x^2 − y\,x − 1/2\,y^2$

> norm(v, 2);   Calculates the length of the vector v .[†]

    v := vector([1, −2, 3]);

    norm(v, 2);   # gives $14^{1/2}$

> plot3d(f, x = a..b, y = c..d);   Plots $z = f(x, y)$ for $a \le x \le b$ and $c \le y \le d$. You can modify and print the plot using menus on plot window.

plot3d(sin(x + y), x = −1..1, y = −1..1);

plot3d([r*sin(s), cos(s), r + s], r = 0..1, s = −Pi..Pi);
# This is used when surface given parametrically.

plot3d(cos(x + y), x = −12..12, y = −12..12, grid = [35, 35], orientation = [85, 30], axis = BOXED, title = `Example`);

plot3d([t*s, exp(s), s], t = −2..1, s = 0..3, view = [−1..1, 0..3, −1..2]);

> taylor(f, x = a, n);   Computes the Taylor series up to degree $n − 1$ of f with respect to x about the point $x = a$. The defaults are $a = 0$ and $n = 6$.

taylor(sin(x));   # gives $x − 1/6 \, x^3 + 1/120 \, x^5 + O(x^6)$

taylor(exp(x), x, 4);   # gives $1 + x + 1/2 \, x^2 + 1/6 \, x^3 + O(x^4)$

taylor(ln(x), x = 1, 3);   # gives $x − 1 − 1/2 \, (x − 1)^2 + O((x − 1)^3)$

> vector([x1, ..., xn]);   Defines vector with n elements.

v := vector([5, 4, 6, 3]);

v[2];   # gives 4

---

3. Stopping and then Restarting a Maple Session

Most students find it invaluable to be able to stop a Maple session and then come back later and restart it as if they had never left. Unfortunately, how to do this depends on the type of machine you are using. Here we describe how this is done when working within X-windows and also when using the Macintosh version of Maple. If you are using a different system you should still find the following helpful in suggesting how it is done on your computer.

One last comment: It is wise to save your session from time to time if your computer system has a tendency to crash unexpectedly.

X-WINDOWS

To save (then quit):

- Select the **Save Text** entry from the **Utilities** menu at the top of the Maple window to save a log of your session. You need to give a name to the file you are copying the session into. Suppose we give it the name **session1**. After entering the name, hit the return key. By the way, only use letters and numbers in the name of the file (do not include characters like . or ;).

- Now use the Maple **save** command to save the session as a special Maple file. The command is as follows:

  > **save** `session1.m`;

  Note the file name is in back quotes and there is a '.m' at the end of the filename.

- Quit Maple. This is optional but anything you do in your Maple session after this will not be saved, unless, of course, you go through the preceding steps again. Also, it goes without saying that you will need to remember the names of the two files you created.

To restart:

- Start Maple.

- Select the **Include File** entry from the **Utilities** menu to retrieve the log of your previous session. You will have to enter the file name where your log is stored. In the preceding example this file was named **session1**. After entering the name, hit the return key.

- Use Maple's **read** command to read in the special Maple file you saved. The command is as follows:

  > **read** `session1.m`;

  Again, note the back quotes.

- You can now continue using Maple as though your original session had not been interrupted. Note, however, that Maple does not redraw the plots when you restart.

MACINTOSH

To save (then quit):

- Select **Save As...** from the **File** menu to save a log of your session. You need to give a name to the file you are copying the session into. Suppose we give it the name **session1**. After entering the name, hit the return key or click **Save**. You should only use letters and numbers in the filename (do not include characters like . or ;). It is always a good idea to save your file periodically as you work (say, every 10 or 15 minutes). To do this, select **Save** from the **File** menu (or press ⌘–S) once the file has been created.

- Quit Maple. Select **Quit** from the **File** menu (or press ⌘–Q). If any changes have been made to the file since you last saved it, Maple will ask you if you would like to save the changes before quitting.

To restart :

- To return to **session1**, double-click on the **session1** icon. This will automatically launch Maple and open a window entitled **Worksheet "session1"**.

- Select **Select All** from the **Edit** menu (or press ⌘–A), and all the text will become highlighted. Press the enter key and every command in the file will be executed from the beginning. *Note:* If you hit the return key instead of the enter key, everything in the window will disappear. If this happens, you can retrieve the text by selecting **Undo Typing** from the **Edit** menu (or press ⌘–Z).

- You can now continue using Maple where you left off.

---

4. Editing Commands in a Maple Session

Because Maple is used interactively, the ability to edit commands and then reissue them is a skill worth acquiring. To explain, suppose you issue the command

> plot(x*sin(x), x = 0..2*Pi);

After looking at the plot suppose you decide the interval $0 \le x \le 2\pi$ is too large and it would be better to use $\pi/2 \le x \le 3\pi/2$. Well, you can either (1) retype a completely new plot command, (2) copy the old command, paste it onto the new line, and then edit it, or (3) move the cursor up to the old command and edit it there. In the case of the last option, if you go back more than one command, you need to be careful about hidden variables (see Section 6). It is recommended that you experiment using (2) and (3), because you will find them very useful when doing the Labs.

5. Idiosyncrasies and Features of Maple

As for any computer program, there are certain aspects of Maple that take some getting used to. For example, when should you use forward quotes and when should you use backward quotes? There are also questions, such as "do I want to use `fsolve` here or would I really be better off with `solve`?" Well, some of these issues are discussed here, but again you should also consider using Maple's on-line help.

`fsolve` vs. `solve`

The `solve` command will solve equations with unevaluated constants (e.g., $x^2 = 2k$) and it will, in some cases, find all of the solutions (real and complex). This is good, but this also limits what it can solve (actually, in some ways, it is limited by what equations mathematicians have been able to find formulas for). On the other hand, `fsolve` will solve many different types of equations but in this case there can be no arbitrary constants in the equation (e.g., it will solve $x^2 = 2$ but not $x^2 = 2k$ unless k has been specified).

There is also an important difference between these commands even for equations they both will solve. This is that `solve` will give the exact solutions to an equation such as $x^2 = 2$ (i.e., it will find the two solutions $\pm 2^{1/2}$) whereas `fsolve` will give decimal solutions (e.g., ± 1.414213562). The latter are not exact but are very accurate approximations. The fact that they are not exact, however, brings us to the next topic.

`evalf,` `fsolve,` and `Digits`

There are certain things we are forced to worry about even if we don't want to, and one example of this is roundoff error. Anytime you tell a computer to try to evaluate something numerically (say, $\sqrt{2}$), it can only use a finite number of digits. The difference between the exact result and the numerically evaluated quantity is known as roundoff error. If this seems to be causing a problem you can increase the value of `Digits,` although you should be aware that this can increase the computing time.

You may now be wondering how you will be able to tell if you should be worrying about roundoff error. Well, there is no way to answer this that will cover all situations, but if you are suspicious of the accuracy of the answer just increase the value of `Digits` to see if the answer is affected. For example, suppose you get an answer like $1 + 0.2 \times 10^{-10}I$. If you think the answer should be real valued, then you would be suspicious of this result. You might also be suspicious because Maple normally keeps 10 digits and this answer says the solution is complex but it is the eleventh digit where the I appears. In either case it would be worth increasing `Digits` to see what happens to the result.

One other thing, if you believe roundoff may be a problem try entering the numbers as fractions rather than in decimal form (e.g., use 13/10 instead of 1.3). This is because in certain situations fractions are treated differently than numbers in decimal form.

`:=` vs. `->`

In Maple you can enter $f(x) = x^2$ as f := x^2 or as f := x->x^2 . The first is called an expression and the second a "function." In this book we will usually use the expression form whenever possible because it is easier. Note that if you use the expression form then you refer to the formula using just f, whereas for the "function" form you must refer to it as f(x) . To illustrate the differences we have the following:

Definition:	f := x^3 + 1;	f := x -> x^3 + 1;
Evaluate at x = 2:	subs(x = 2, f);	f(2);
Differentiate:	diff(f, x);	diff(f(x), x);
Integrate:	int(f, x = 0..1);	int(f(x), x = 0..1);

`;` vs. `:`

If you want Maple to display the response so you can see it, use a semicolon (;) after the command. If you don't want to see the response (e.g., maybe it is so long that it is unintelligible), then use a colon (:) after the command.

Back Quotes vs. Quotes

Back quotes are used to identify text to go into a command, such as the title for a plot (e.g., title = `Plot`). Quotes are used, among other things, to return a variable to unevaluated status and to undefine a "function" (e.g., x := 'x').

sqrt(·)

One source of potential trouble is the square root function. To illustrate, if you enter the expression sqrt(f^2) then Maple will return f . This is OK, but what if f is negative and you expect sqrt(f^2) to be positive? For example, Maple states that sqrt(sin(x)^2) = sin(x) . This is positive if sin(x) is positive and negative if sin(x) is negative. A simple way to make sure the result is nonnegative is to use something like abs(sqrt(f^2)) . However, it is recommended that you do **not** do this unless absolutely necessary. The reason is that integration formulas and solutions for equations are limited if the abs(·) function is present.

I, Pi, O, etc.

Like any good computing system, Maple has a certain collection of well used mathematical constants and functions available. These have special symbols and you cannot use them for other tasks. So, for example, Maple reserves I to represent $\sqrt{-1}$ and Pi for π . You can get a listing of these special symbols by using the help commands ?I or ?Pi .

Do Not Do This

Because Maple is capable of solving some very difficult mathematical problems, it has been equipped with numerous special functions and constants. Examples are the trigonometric functions and π. Needless to say, things can really go haywire if you attempt to set them to other values, e.g., if you let Pi := 0. Some of these (like π and the trigonometric functions) are used so much that you probably will not set them to other values. However, there are others to watch for and they include: max, min, length, and sum. Others are left, right, and list. In other words, **never** try to define these as they are already defined by Maple. You may use, if you wish, slight variations, such as Max, Min, or MAX, MIN, or max1, min1.

---

6. A Common Error

It is inevitable that you will occasionally have trouble getting your ideas across to Maple, and this can quickly lead to a very frustrating situation. It is impossible to anticipate every possibility, but there is one that is quite common. This is that a variable has already been assigned, and you try to use it in another context. This causes almost everyone problems at one time or another, because after working on a Maple session for an hour or two it is very easy to forget what variables you have used.

To illustrate this, consider the following short Maple session...

> f := x^3 + 1;

$$f := x^3 + 1$$

> x := 1;

$$x := 1$$

> f;

$$2$$

> plot(f, x = 0..1);

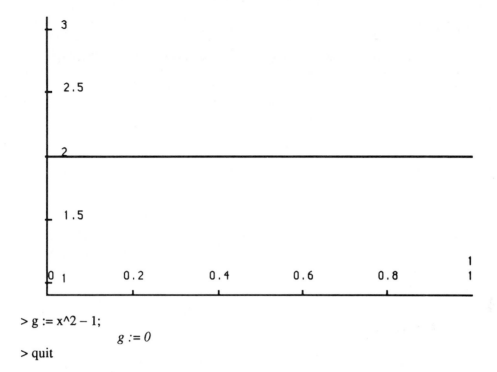

```
> g := x^2 - 1;
                    g := 0
> quit
```

The assignment x := 1 appearing in this session has a profound effect on the last two commands. In the plot, we get the constant (f = 2) and not the original cubic. Also, in the last command, g is not a quadratic but simply a constant. If this is what was desired, then fine. However, if you expected to see a cubic in the plot and a quadratic for g , then x has to be unassigned its earlier value. This is accomplished by issuing the command (before the plot command)

```
> x := 'x';
```

There is another way assignment can occur that can cause the same problems illustrated in the preceding example. When using a do-loop the index, or counter, leaves the loop with an assigned value. An example of this is given in the solution guidelines for Problem 2 of Laboratory 4.

If you are not sure whether or not you have assigned a value to a variable, you can always check. For example, for the variable x you issue the command

```
> x;
```

If the reply is just x := x , then it is unassigned. This simple test will work unless x has been defined earlier to be a vector or an array. In this case you can get Maple to tell you about x by using the command

```
> print(x);
```

This also works to find out if x has been given a numerical value (such as 6).

---

7. Maple References

Maple® V Flight Manual: Tutorials for Calculus, Linear Algebra, and Differential Equations, by W. Ellis, E. W. Johnson, E. Lodi, and D. Schwalbe, Brooks/Cole Publishers, Pacific Grove, CA, 1992.

First Leaves—A Tutorial Introduction to Maple V, by B. W. Char, K. O. Geddes, G. H. Gonnet, B. L. Leong, M. B. Monagan, and S. M. Watt, Springer-Verlag, New York, 1991.

Maple V Library Reference Manual, by B. W. Char, K. O. Geddes, G. H. Gonnet, B. L. Leong, M. B. Monagan, and S. M. Watt, Springer-Verlag, New York, 1991.

Maple V Language Reference Manual, by B. W. Char, K. O. Geddes, G. H. Gonnet, B. L. Leong, M. B. Monagan, and S. M. Watt, Springer-Verlag, New York, 1991.

Part 3

Sample Laboratory

The following is a short Laboratory, and its solutions, to illustrate what might typically be expected when writing up the solutions. Of course, your instructor will tell you in more detail what is expected in your own class.

In the solutions, the command `printplot` is used. This is not a regular Maple command but a procedure that was written to print two-dimensional plots. The commands in the procedure can be found in Appendix A2.

0

Limits and Continuity

Objective

To use Maple to evaluate limits.

One of the fundamental concepts in calculus is that of a limit. In this Lab you are asked to use Maple to find limits of relatively complicated functions. You can also use Maple to prove what the value of the limit is, and this is the subject of Laboratory 2.

EXAMPLE

A typical (harder) homework problem in calculus involving limits is the following: As $x \to 0$, find the limit of

$$f(x) = \frac{\text{sqrt}(9 + 5 * x) - \text{sqrt}(9 - x)}{x} \quad .$$

Maple can determine the limit very easily, but first we will examine the function to see if we can anticipate what the limit is. The first step is to enter the formula for f into Maple as follows:

> f := (sqrt(9 + 5*x) − sqrt(9 − x)) / x;

Our approach is to plot $f(x)$ over an interval that contains $x = 0$. The default interval for plotting in Maple is $-10 < x < 10$, but this is too big, because of the square roots in the definition of f. In fact, we need to restrict x so that $-9/5 < x < 9$.

> plot(f, x = −9/5..9);

It appears that the limit is 1, but to check we'll zoom in on $x = 0$ by reducing the size of the x interval. We'll also restrict the range along the y-axis (to a smaller neighborhood of $y = 1$), and we'll add a title to the plot.

> plot(f , x = −1/2..1/2, y = 1/2..3/2, title = `Plot for Example`);

We have made the x interval so small that the function appears to be a straight line. What we see is that this new plot supports our hypothesis that the limit is one. To verify this we use the `limit` command to have Maple calculate the limit:

> limit(f, x = 0);

Note that this command requires the function as well as the limit point. Remember that if you are unsure what information a command needs, you can always use the on-line help facility. In any case we have found that, indeed, the limit is 1.

Laboratory Problems

1. Let

$$f(x) = x^3 + \frac{\sin(2x)}{x} .$$

   (a) Plot this function, and from the plot estimate the value of the limit as $x \to 0$. Hand in a plot that supports your answer.

   (b) Check on your estimate by having Maple evaluate the limit.

2. Let

$$f(x) = \frac{x - k}{2x^5 - x^4 + 2x^2 + x - 1} .$$

   (a) For what value of k is $f(x)$ continuous for all x?

   (b) Find the limit of f as $x \to k$ for the value of k found in (a).

3. Let

$$f(x) = \cos(\frac{1}{x}) \ .$$

(a) Maple gives an interval for the limit of this function as $x \to 0$. Find this interval.

(b) Plot f, and from this explain where Maple gets this answer. Also explain why, based on the definition of a limit given in your textbook, this answer is not correct. What is happening here is that Maple is using a generalization of the limit definition.

LABORATORY SOLUTIONS

```
    | \ ^ / |          MAPLE V
._ | \ |     | / |_ .  Copyright (c) 1981–1990 by the University of Waterloo.
  \    MAPLE    /       All rights reserved.  MAPLE is a registered trademark of
  <_____ _____>         Waterloo Maple Software.
        |               Type ? for help.
>
># 			LABORATORY 0 SOLUTIONS
>
># PROBLEM 1
># The first step is to enter the formula for f:
> f := x^3 + sin(2*x)/x;
                        3    sin(2 x)
                f := x    +  --------
                                x


># Part (a)
># It is not clear what interval to use here so we use the default
># (of −10 < x < 10)
> plot(f);

># The interval is too large to estimate the value at  x = 0,  so let's shrink it.
> plot(f, x = −1..1);

># It looks like the limit is 2, but let's reshrink to make sure and add
># a title for the write-up:
> plot(f, x = −0.1..0.1, y = 1.5..2.5, title = `Plot for Problem 1`);
> printplot(f, x = −0.1..0.1, y = 1.5..2.5,title = `Plot for Problem 1`);

># This plot supports our hypothesis that the limit is 2 .
```

```
># Part (b)
> limit(f, x = 0);
                            2
```

```
># So, we were correct!
>
>
># PROBLEM 2
># Part (a)
># This is a rational function and is therefore continuous everywhere except at
># points where the denominator is zero.  The easiest way to get an idea of
># where these points are is to plot it:
> f := (x − k)/(2*x^5 − x^4 + 2*x^2 + x − 1);
```

$$f := \frac{x - k}{2x^5 - x^4 + 2x^2 + x - 1}$$

```
> bottom := 2*x^5 − x^4 + 2*x^2 + x − 1;
```

$$bottom := 2x^5 - x^4 + 2x^2 + x - 1$$

```
> plot(bottom);
```

```
># From this plot it is apparent that the denominator takes on both positive
># and negative values. This means (since the denominator is continuous) that
># it must go through zero at least once. The scale in the graph is too large
># to estimate where this occurs, so we reduce it. Note from the plot that
># we can restrict our attention to −5 < x < 5 .
> plot(bottom, x = −5..5);
```

```
># The x interval needs to be reduced further...
> plot(bottom, x = −1..1);
```

```
># Well, it appears that the only place bottom is zero is at  x = 1/2 .  Let's
># check on this using fsolve:
> fsolve(bottom = 0, x, x  = −1..1);
                            .5000000000
># Good, we were right.
># Now, in our formula for  f  we do not want to divide by zero.  To compensate
># for this we will choose  k  so the numerator is also zero at  x = 1/2 .
```

```
> k := 1/2;
```
$$k \ := \ 1/2$$

```
> f;
```
$$\frac{x \ - \ 1/2}{2 \ x^5 \ - \ x^4 \ + \ 2 \ x^2 \ + \ x \ - \ 1}$$

```
># It remains to see that the limit exists:
> limit(f, x = 1/2);
```
$$8/25$$

```
>
>
># PROBLEM 3
> f := cos(1/x);
```
$$f \ := \ \cos(1/x)$$

```
># Part (a)
> limit( f, x = 0);
```
$$-1 \ .. \ 1$$

```
># The interval is −1 =< x =< 1 .
># Part (b)
> plot(f);

># It's unclear what's happening near  x = 0, so let's shrink the interval:
> plot(f, x = −1..1);

># Let's reduce the interval a little more, and add a title for the write-up:
> plot( f, x = −0.1..0.1, −1.1..1.2, title = `Plot for Part 3(b)`);
> printplot( f, x = −0.1..0.1, −1.1..1.2, title = `Plot for Part 3(b)`);

># Maple gets its interval from the fact that this function is wildly
># oscillating between −1 and 1 near x = 0 . This answer is incorrect,
># using the definition of a limit given in the textbook, because the value
># of a limit must be unique. Therefore, this function has no limit as x −> 0
># because of its rapid oscillations.
> quit
```

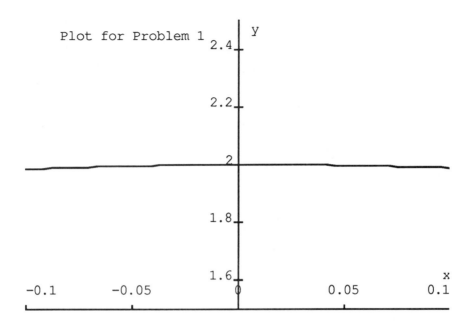

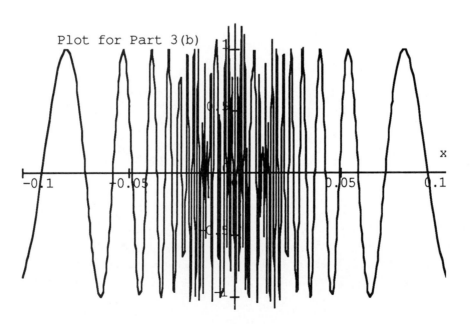

Part 4

Laboratories

1

Understanding Absolute Values

Objective
To introduce you to Maple and how to use it to develop new approaches to problem solving.

The purpose of this Lab is to help you get started in learning about Maple and about how it can be used as a problem solving tool in calculus. Several sample example problems are given that involve absolute values, inequalities, functions, and graphs. Following the statement of these problems are their detailed solutions worked out using Maple. You should work through these exercises and become familiar with the Maple commands that are used. Because this is an introductory Lab, we have provided lots of details that will help you in using Maple effectively.

EXAMPLE 1
Let $f(x) = x^4 - 3x^3 - 2x^2 - 9$.

(a) Find all values of x where $f(x) = 0$.

(b) Graph the curve $y = f(x)$ over the interval $[-5, 5]$.

(c) On the same graph, plot the function $y = f(x)$ and the straight line
$y = -2x + 10$.

(d) Find the x- and y-coordinates of all points where the straight line intersects the curve.

EXAMPLE 2

(a) Find all values of x that satisfy $|5x + 7| = 3$.

(b) Find all values of x that satisfy $|2x - 3| \le 5$.

(c) Find all values of x that satisfy $|2 - 1/x| > 3$.

The following solutions lead you through these sample problems in a rather detailed way. Remember that the goal here is to introduce you to some Maple commands and to illustrate that having Maple as a tool permits new approaches to problem solving.

Solution to Example 1   Enter the formula for f by using the following command:

> f := x**4 – 3*x**3 – 2*x**2 – 9;

$$f := x^4 - 3x^3 - 2x^2 - 9$$

Notice that := is used to assign the name f to the formula. Also notice that the command ends with a semicolon. Every Maple command must end with a semicolon or colon. We have used x**4 to indicate x^4, but we could have used x^4 instead.

(a) To find the roots of $f(x) = 0$, we issue the command

> fsolve(f = 0, x);

$$-1.435531110, 3.714140442$$

Maple returns a list of two real roots separated by a comma. The other two roots happen to be complex. For a polynomial equation such as the one in this example, the `fsolve` command will return all real roots.

In order to refer to the first or second root in the preceding list, we need to give it a name. For example, the following command gives the name "soln" to the list of roots:

> soln := fsolve(f = 0, x);

$$soln := -1.435531110, 3.714140442$$

Maple then will recognize the first root in the list as soln[1] and the second root as soln[2] .

```
> soln[1];
```
$$-1.435531110$$
```
> soln[2];
```
$$3.714140442$$

(b) To sketch a graph of the f over the interval $[-5, 5]$, use the plot command:

```
> plot( f, x = -5..5);
```

It is a little hard to see the details of this graph. We can limit the range on the y-axis to be from -50 to 100 by using the command

```
> plot( f, x = -5..5, y = -50..100);
```

Another way to get a better plot is to restrict the domain of the x values to an interval where we are interested in seeing more detail. For example, the following gives a nice plot:

```
> plot( f, x = -2..4, title = `A Nice Plot`);
```

(c) Maple can plot several graphs on the same axes. Define the function $g(x) = -2x + 10$ by entering the command

```
> g := -2*x + 10;
```

Then enter the following command to plot both f and g on the same axes:

```
> plot( {f, g} x  = -2..4, title = `Plots of f and g`);
```

(d) We can estimate the points of intersection from this plot. Better estimates can be obtained by replotting the functions using a smaller interval (e.g., x = -2.. -1.5 and x = 3.5..4). However, to obtain more precise values of the x-coordinates of the intersection points, we use the `fsolve` command:

```
> fsolve( f = g, x);
```

Maple returns two x values in a list. Again it would have been better if we had named the list. So let's use `fsolve` again.

```
> s := fsolve( f = g, x);
```
$$s := -1.824622361, 3.751039107$$

Enter the Maple commands that will define x1 and x2 as the first and second entries in this list s :

> x1 := s[1];

$$x1 := -1.824622361$$

> x2 := s[2];

$$x2 := 3.751039107$$

To obtain the corresponding y-coordinates of these points of intersection, we use the substitute command:

> y1 := subs(x = x1, f); y2 := subs(x = x2, f);

$$y1 := 13.64924472$$

$$y2 := 2.49792174$$

Notice that you can enter more than one Maple command on a single line.

Solution to Example 2   Maple uses abs(u) instead of $|u|$ to refer to the absolute value of u .

(a) To find all x so that $|5x + 7| = 3$, use the solve command:

> solve(abs(5*x + 7) = 3, x);

$$-4/5, -2$$

If we want to find a decimal number for $-4/5$ we can use the `evalf` command. Of course, we all know that $-4/5 = -0.8$, but let's illustrate this command:

> evalf(-4/5);

$$-.8000000000$$

As another example,

> evalf(2*Pi + sqrt(7));

$$8.928936619$$

Notice how $\pi = $ Pi is entered into Maple.

Remember that $|u| = u$ if u is nonnegative, and $|u| = -u$ if u is nonpositive. So to get one root Maple solves the equation

> solve(5*x + 7 = 3, x);

$$-4/5$$

and to get the other root Maple solves the equation

> solve(−(5*x + 7) = 3, x);

$$-2$$

The `solve` command returns an exact solution to an equation if one can be found. That is, the `solve` command works very much like we do when we use pencil and paper to solve an equation. We also can use the command `fsolve,` which finds solutions to equations using a numerical algorithm. For example,

> fsolve(abs(5*x + 7) = 3, x);

$$-.8000000000$$

Using this command, Maple returns a 10-digit decimal for one of the roots but misses the other root, -2. For functions that are not polynomials `fsolve` attempts to compute only one real root. However, there is an option that allows us to specify an interval on the x-axis in which to look for a root. For example, to look in the interval $-10 \le x \le -1$, use the command

> fsolve(abs(5*x + 7) = 3, x, −10..−1);

$$-2.000000000$$

(b) To find all x so that $|2x - 3| \le 5$, use the command

> solve(abs(2*x − 3) <= 5, x);

$$\{-1 <= x, x <= 4\}$$

Note the way we have entered the $\le$ sign in this command. Also note the solution, which states that $-1 \le x \le 4$. The `fsolve` command cannot deal with inequalities that call for a solution that is an interval. Remember, `fsolve` returns a decimal number.

(c) To find all x so that $|2 - 1/x| > 3$, enter the command

> solve(abs(2 − 1/x) > 3, x);

Error, (in series/leadterm) unable to compute leading term

Does this mean that there are no values that satisfy this inequality? No, for example, if $x = -1/10$ then $|2 - 1/x| = 12 > 3$. **Maple is just confused!** It is important to realize that Maple cannot do everything, and in this example the procedure Maple uses to solve this type of problem does not work (hence the error message).

This inequality is not hard to solve using pencil and paper. If $2 - 1/x > 0$, then $|2 - 1/x| = 2 - 1/x$, and so $2 - 1/x > 3$ implies $-1/x > 1$. Therefore x cannot be positive, because this would contradict $-1/x > 1$. So multiplying the inequality $-1/x > 1$ by $x < 0$ reverses the inequality to give $-1 < x$. If $2 - 1/x < 0$, then $|2 - 1/x| = -(2 - 1/x)$, and a similar argument shows that $-(2 - 1/x) > 3$ implies that $x < 1/5$. Therefore $|2 - 1/x| > 3$ in the interval $-1 < x < 1/5$.

The preceding derivation is the solution approach we would use with pencil and paper. But if the function inside the absolute value signs were very complicated we would be doomed to failure in trying to solve the inequality analytically. Maple allows us to think of alternative approaches to solving such a problem. For example, why not just plot a graph of $|2 - 1/x|$ and see where it is greater than 3?

```
> f := abs(2 − 1/x);
> plot( f, x = −3..3);
```

This plot is not much help, because when x is near zero the values of f are extremely large. Let's limit the vertical range of the plot:

```
> plot( f, x = −3..3, 0..5);
```

To help see where $f > 3$, it helps to plot the constant $g = 3$ on the same axes:

```
> g := 3;
> plot( {f, g}, x = −3..3, 0..5);
```

We can now use the `fsolve` command to find where $f > 3$:

```
> fsolve(f = g, x);
```

$$fsolve(abs(2 - 1/x) = 3, x)$$

Maple returns the expression we entered and seemingly cannot solve the problem. But this is because Maple needs to know whether x is positive or negative in order to determine a numerical value for $abs(2 - 1/x)$. So, we limit the interval over which Maple searches for a solution:

```
> x1 := fsolve( f = g, x, −2..0);
```

$$x1 := -1.000000000$$

```
> x2 := fsolve( f = g, x, 0..1);
```

$$x2 := .2000000000$$

Therefore, the interval of x values that satisfies the inequality is $-1 < x < 2$. This agrees with the solution we have obtained before.

Laboratory Problems

In this Lab you are to carry out steps similar to those in the examples. The three problems here are independent of one another.

1. Let $f(x) = 2 - x$ and $g(x) = x^3 - 4x^2 + 3x + 3$. Also, let $h(x) = |f(x)/g(x)|$.

    (a) Plot the graphs of f and g on the same axes over the interval $[-2, 4]$. Find the x- and y-coordinates of any intersection points of f and g. Also, explain why the interval $-2 < x < 4$ is adequate to determine all of the intersection points.

    (b) Find all the real values of x where $g(x) = 0$.

    (c) Find all the real values of x where $h(x)$ is not defined.

    (d) Find all values of x where $h(x) < 1$.

2. Find all values of x where $|2x - 3/(x^3 + 2x + 3)| < 2$.

3. Maple knows many special functions including the trigonometric functions.

    (a) Enter $\sin(x)$ into Maple by using the command

    > f := sin(x);

    Letting $g(x) = 1 - x^2$, find the x- and y-coordinates of all intersection points where $f = g$.

    (b) Find all values of x where $|\sin(x) - 2\cos(x)| < 1$ in the interval $[0, \pi]$.

2

The Cast of Characters

Objective
To use Maple to visualize and gain understanding of some of the most important functions in calculus.

Functions play a vital role in calculus. In this Lab we will look at many of the functions that we will eventually see in our calculus course. We will use Maple to investigate what the graphs of these functions look like and to investigate relationships between some of the functions.

For example, the function f that maps the real number x to x^2 is usually indicated by $f(x) = x^2$. Here the domain of f is the set of all real numbers, and the range of f is the set of nonnegative real numbers. In Maple one way to define this function f is to use the command

> f:= x –> x**2;

$$f := x \to x^2$$

This arrow notation emphasizes the fact that f is a rule that maps x to x^2. Also, this way of entering f into Maple allows us to obtain values $f(x)$ simply by entering commands such as

> f(2);

$$4$$

> f(Pi);

$$Pi^2$$

> f(2+ h);

$$(2 + h)^2$$

To get a decimal number for Pi^2 , we use the `evalf` command.

> evalf(f(Pi));

$$9.869604404$$

Linear Functions

Linear functions get their name from the fact that their graphs are straight lines. Any linear function can be defined by

> f := x −> a*x + b;

$$f := x -> a\,x + b$$

where a and b are real numbers. The y-intercept of the graph of f is b as we see from calculating

> f(0);

$$b$$

The graph of f also intercepts the x-axis at the point where $f(x) = 0$.

> solve(f(x) = 0, x);

$$- b/a$$

Of course, if $a = 0$ then the graph is parallel to the x-axis with no intercept. Notice that in the Maple commands we refer to the function as $f(x)$.

Laboratory Problem

1. (a) Use Maple to plot the linear function $f(x) = Pi*x + sqrt(Pi)$, and find the x- and y-intercepts. Enter $f(x)$ into Maple using the preceding syntax.

    (b) Calculate $f(1.3)$, $f(5*sqrt(3))$, and $f(7)$ as decimal numbers.

Polynomial Functions

Polynomial functions are functions of the form

$$f(x) = a_0 + a_1x + a_2x^2 + a_3x^3 + ... + a_nx^n \, ,$$

where a_0, a_1, a_2, a_3, ... , a_n are real numbers. The nonnegative integer n is called the degree of the polynomial. For example,

> f := x -> 3 + 2.5*x - sqrt(5)*x**2 + 2*x**3;

$$f := x -> 3 + 2.5\, x - sqrt(5)\, x^2 + 2\, x^3$$

defines a third-degree polynomial function. The function g that follows is also a third degree polynomial function:

> g := x -> 2*x**3;

$$g := x -> 2\, x^3$$

Laboratory Problems

2. (a) Notice that the function g is just the highest power term in the polynomial function f. Calculate decimal values for f and g evaluated at $x = 1$, 10, 100, and 1000.

    (b) If you rounded off the values $f(1000)$ and $g(1000)$ to four digits following the decimal point, would you get the same value?

    (c) Plot $f(x)$ and $g(x)$ on the same axes for $x = -1000..1000$. The graphs for f and g look very much alike. This happens because for large values of x , the highest degree term in f(x) dominates the other terms. We say that for large values of x , f(x) behaves like its highest power term.

(d) For values of x near 0 , is g(x) a reasonable approximation to f(x) ? Explain.

3.   We saw earlier that for large x a polynomial behaves like its highest power term. For each of the following polynomials p , what happens to the values of p(x) as x approaches ∞ , and what happens to p(x) as x approaches − ∞ ?

(a) p(x) = 2 + 3*x + 5*x**3

(b) p(x) = −5 − 3*x + 4*x**2 + 3*x**3 + x**4

(c) p(x) = 2 + 5*x + 2*x**3 − 4*x**6

Using Maple it is easy to obtain the real roots of most polynomial functions. For example, to find the roots of

> f := x −> 3*x**7 + 5*x**6 − 4*x**3 + 2*x + 1;

$$f := x -> 3\,x^7 + 5\,x^6 - 4\,x^3 + 2\,x + 1$$

we could try the solve command:

> solve(f(x) = 0, x);

$$RootOf(3\,\_Z^7 + 5\,\_Z^6 - 4\,\_Z^3 + 2\,\_Z + 1)$$

This response means that Maple cannot find a solution analytically, so we can try numerically by using

> fsolve(f(x) = 0, x);

$$-1.853737614$$

For polynomials, Maple will try to find all real roots. The current f(x) has only one real root as we can see by plotting:

> plot(f(x), x = −5..5);

This plot is not very revealing, so we restrict the range in the next plot.

> plot(f(x), x = −5..5, −50..50);

Laboratory Problem

4. Find the real roots for each of the three polynomial functions given in Problem 3.

Rational Functions

Rational functions are just quotients of polynomial functions. For example, consider the rational function

> f := x −> (x**2 + 1)/x;

$$f := x \rightarrow \frac{x^2 + 1}{x}$$

Clearly, f is not defined at $x = 0$ because the denominator is zero there. For large x , the numerator behaves like x^2 and the denominator behaves like x , so f must behave like $x^2/x = x$. Let's check this by plotting f(x) and g(x) = x on the same axes:

> g := x −> x;

$$g := x \rightarrow x$$

> plot({f(x), g(x)}, x = −5..5, −5..5);

Notice that we restricted the range of the plot to −5..5 to get a better picture.

Laboratory Problem

5. Let $f(x) = 3x^4 - 5x^2 + 3x + 4$, $g(x) = 2x^2 + 4$, and define the rational function $h(x) = f(x)/g(x)$.

    (a) Is h defined for all values of x ? Explain.

(b) Find a function k(x) so that, for large values of x , the graphs of
h(x) and k(x) look very similar.

Exponential Functions

An exponential function has the form

$$f(x) = a^x,$$

where a is a positive constant called the base. To investigate the behavior of this
function, let's take the particular case of a = 2 :

> f := x -> 2**x;

$$f := x \rightarrow 2^x$$

How does this function compare, say, with the case of when a = 3 ? To find out
let's look at the plots of a^x when a = 2 and when a = 3 :

> plot({2**x, 3**x}, x = -5..5, 0..250);

As we expected 3^x grows much faster than 2^x .
At this point it is helpful to review the

Rules of exponents:
If a > 0 and b > 0 , and x and y are real numbers, then

$$a^x b^x = (ab)^x, \ a^x a^y = a^{x+y}, \ (a^x)^y = a^y, \ \frac{a^x}{a^y} = a^{x-y}.$$

We saw that 3^x grows much faster than 2^x as x increases. To get some idea of
how fast $f(x) = 3^x$ grows, we can use the rules of exponents to calculate

$$\frac{\text{rise}}{\text{run}} = \frac{f(x+h) - f(x)}{h} = \frac{3^{x+h} - 3^x}{h} = \frac{3^x(3^h - 1)}{h}$$

$$= 3^x \frac{3^h - 1}{h}$$

$$= 3^x g(h),$$

where $g(h) = (3^h - 1)/h$. Defining g in Maple

> g := h -> (3**h - 1)/h;

$$g := h -> \frac{3^h - 1}{h}$$

we can calculate $g(h)$ for small values of h.

> evalf(g(0.01));   evalf(g(0.001));   evalf(g(0.0001));

1.10466919
1.0992160
1.098673

Thus, for the function 3^x,

$$\frac{\text{rise}}{\text{run}} \quad \text{approaches} \quad 1.099*(3^x) \quad \text{as } h \text{ approaches zero.}$$

Laboratory Problem

6. Use a similar analysis to show that for the function 2**x ,

$$\frac{\text{rise}}{\text{run}} \quad \text{approaches} \quad 0.693*(2^x) \quad \text{as } h \text{ approaches zero.}$$

We could have used any base a in the preceding analysis instead of just the bases 2 and 3. Thus, for any exponential function a^x , the ratio rise/run approaches (some constant)*a^x as the run h approaches zero. Further, we know from the derivation already given that the constant is just the limiting value of the ratio

$$\frac{a^h - 1}{h} \quad \text{as } h \text{ approaches zero.}$$

What must the base be so that this constant is one? One thing to notice is that as the base increases so does the constant. For example, when the base is 2 the constant is 0.693 and when the base is 3 the constant is 1.099. Thus there must be some

some number, let's call it e , between 2 and 3, for which the constant is exactly equal to 1. This means that we have defined e to be the base so that for the function e^x,

$$\frac{\text{rise}}{\text{run}} \text{ approaches } 1*(e^x) \text{ as } h \text{ approaches zero.}$$

Laboratory Problem

7. (a) Fix h = 0.0001 and define

$$f(a) = \frac{a^h - 1}{h} \quad .$$

   Find the value of a so that f(a) =1 by using the `fsolve` command.

   (b) Now fix h = 0.00001 and find the value of a so that f(a) = 1 .

   (c) Given the results from (a) and (b), what must the value of e be to three decimal places?

The base e is an extremely important base, and

The exponential function e^x is one of the most important functions in all of calculus.

Maple uses the notation exp(x) to refer to the function e^x . For example, to see a graph of exp(x) , we use the command

> plot(exp(x), x = −1..3, title = `A Plot of the Exponential Function`);

You can also have Maple calculate the value of e using the command

> evalf(exp(1));

2.718281828

As expected, e lies between 2 and 3 . To get a more accurate approximation of the value of e we can increase the value of `Digits.` For example, suppose we take

> Digits := 50;

and then issue the command again:

> evalf(exp(1));

$$2.7182818284590452353602874713526624977572470937000$$

It is important to realize that, like π and $\sqrt{2}$, the number e is irrational. Before ending this example we should reset `Digits` to its original value:

> Digits := 10;

Logarithms

Most of us have used logarithms to the base 10, and we recall that if log denotes the logarithm to base 10, then $\log(x)$ is defined by

$$\log(x) = y \quad \text{if and only if} \quad x = 10^y .$$

So $\log(x) = y$ means that y is the power to which we raise the base 10 to get x. For example, $\log(100) = 2$ since $100 = 10^2$.

Because the base e is so important, logarithms to the base e are also very useful. They are called natural logarithms (by comparison, those to the base 10 are called common logarithms). We use the notation $\ln(x)$ to designate the logarithm to the base e. Thus $\ln(x)$ is defined by

$$\ln(x) = y \quad \text{if and only if} \quad x = e^y .$$

Maple uses $\log10(x)$ to refer to logarithms to the base 10 and uses $\ln(x)$ to refer to logarithms to the base e.

Laboratory Problem

8. (a) Explain why $\log(x)$ and $\ln(x)$ are only defined when $x > 0$.

   (b) Plot the two functions $\log10(x)$ and $\ln(x)$ over the interval $x = 0..5$.

   (c) Use the definition of $\ln(x)$ given here to show that $\ln(e^x) = x$ and that $e^{\ln(x)} = x$.

Inverse Functions

We say that $\ln(x)$ and e^x are inverses of one another because

$$\ln(e^x) = x \text{ and } e^{\ln(x)} = x.$$

In general, we usually use the notation f^{-1} to denote the inverse of a function f, and we write

$$f^{-1}(y) = x \text{ if and only if } f(x) = y.$$

So, the inverse function f^{-1} maps a value y onto the value x where $f(x) = y$. This means that if there is more than one value of x for which $f(x) = y$, then the inverse function is not defined.

Another way to see that $\ln(x)$ and $\exp(x)$ are inverses of one another is to define the function

> expinverse := y –> fsolve(exp(x) = y, x);

$$expinverse := y \rightarrow fsolve(exp(x) = y, x)$$

We can then plot expinverse using the command

> plot(expinverse, y = 0..5);

Notice that we did not use expinverse(y) in the `plot` command. To get a slightly better plot we restrict the range to $-4.5..2$ with the command

> plot(expinverse, y = 0..5, –4.5..2);

Now we can compare expinverse and the function $\ln(x)$ and actually see that they are identical by plotting them both on the same axes:

> plot({expinverse, ln(y)} , y = 0..5, –4.5..2);

Laboratory Problems

9. (a) Explain why the function $f(x) = 10^x$ has an inverse.

    (b) As in the preceding example, define finverse as the function that

(b) As in the preceding example, define finverse as the function that maps y onto the value of x where $10{**}x = y$. Plot finverse for $y = 0..3$.

(c) Show that finverse and log10 are identical functions by plotting both and seeing that their graphs are identical.

10. (a) Explain why the function $f(x) = x{**}3 + x - 9$ has an inverse.

   (b) Call the inverse function finverse, and plot finverse for $y = -10..10$ and with the range restricted to $-10..10$.

   (c) Plot finverse, $f(y)$, and the function $g(y) = y$ on the same axes for $y = -10..10$ and with the range restricted to $-10..10$.

   (d) What does this graph in part (c) suggest about the relationship of finverse and f?

Trigonometric Functions

The trigonometric functions are certainly important functions. Most of us are familiar with these functions, and Maple uses the usual names to refer to trigonometric functions, namely,

$$\sin(x), \cos(x), \tan(x), \cot(x), \sec(x), \text{ and } \csc(x).$$

Laboratory Problem

11. (a) Use Maple to try and construct a plot of $\tan(x)$ for $x = -2{*}\text{Pi}..2{*}\text{Pi}$. Why does Maple have trouble constructing a reasonable plot over these values of x ?

   (b) Replot $\tan(x)$ over $x = -2{*}\text{Pi}..2{*}\text{Pi}$ but restrict the range to $-5..5$. Does $\tan(x)$ have an inverse function for $x = -2{*}\text{Pi}..2{*}\text{Pi}$? Explain.

   (c) Use the `fsolve` command with the domain of search restricted to $x = -1.5..1.5$ to find a value of x so that $\tan(x) = 10$.

   (d) If we restrict the domain of $\tan(x)$ to $x = -1.5..1.5$, then $\tan(x)$ has an inverse function. Call the inverse function taninverse, and use the `fsolve` command with the domain of search restricted to

$x = -1.5..1.5$ to define taninverse . Plot taninverse for $y = -10..10$. By the way, you should be aware that Maple does know about the inverse of $\tan(x)$ and uses the name arctan instead of taninverse for it.

3

The Limit of a Function

Objective

To enhance your conceptual understanding of the "epsilon-delta" definition of the limit of a function.

In this Laboratory you are asked to do the kind of analysis that is necessary if one uses the formal epsilon-delta definition to explain what it means for $\lim_{x \to a} f(x) = L$. This is an introductory Laboratory, so we provide lots of details on how to use Maple for this type of analysis. We begin with an example that you should work using Maple to familiarize yourself with the problem.

EXAMPLE

A manufacturer has two wires that need to be welded together (see the figure that follows). One wire is straight and runs along the straight line $g(x) = 11$, while the other wire is curved and follows the graph of the function $f(x) = 2x^3 - 4x^2 + 3x + 10$. These two wires cross at the point $(x , y) = (1, 11)$. The x- and y-axes are both measured in inches.

The welding device makes a weld that is a vertical strip 1 inch long, and the center of this strip must lie on the straight wire. Moreover, in order to weld both wires together successfully the weld must overlap both wires.

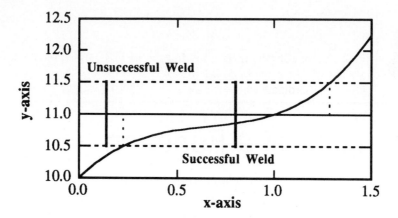

(a) Ideally, the manufacturer would like to place the weld exactly at the point of intersection of the two wires, although this is not necessary for a successful weld. Also, it is quite costly to position the welding device precisely at x = 1 . Find the largest interval about x = 1 in which the x-coordinate of the strip weld can lie and still yield a successful weld.

A GUIDED TOUR THROUGH PART (a)

You will find it helpful to plot f(x) , g(x) , and the lines y = 10.5 and y = 11.5 on the same graph. To do this first define f and g for Maple using the commands:

```
> f := 2*x^3 – 4*x^2 + 3*x + 10;
> g := 11;
```

and then use the `plot` command:

```
> plot ( { f, g, 11.5, 10.5}, x = 0..2, title = `Plot for Part (a)` );
```

The strip weld must be perpendicular to the middle line g = 11 on the plot and extend from the top line y = 11.5 to the bottom line y = 10.5 . A successful weld must also overlap both wires. For example, a weld at x = 0.14   would be unsuccessful but a weld at x = 0.8 would be successful (see the figure above).

To determine how far to the left of x = 1 we can successfully weld, we need to find the x-coordinate where f intersects the bottom line y = 10.5 . To find this x-coordinate, use the `fsolve` command as follows:

> x1 := fsolve(f = 10.5, x);

$$x1 := .2281554937$$

It is convenient to name things in Maple, just in case we want to refer to them later, so here we have named the value of the solution as x1 .

Now find the x-coordinate where the top line y = 11.5 intersects f, and name it x2 . Write the command that will do this in the following space and then enter it into Maple.

>

You should have found x2 = 1.287371537 . Note that x1 and x2 are the endpoints of the largest interval about x = 1 that will yield a successful weld.

(b) Unfortunately, the welding device does not always make a weld that extends exactly 1/2 inch above and below the straight wire. The manufacturer knows precisely how the local conditions (humidity, temperature, etc.) at the time of the weld affect the length of the weld. The manufacturer would like you to find the largest interval about x = 1 that will yield a successful weld when the strip extends a distance ε above and below the straight wire. Specify the interval by giving its endpoints $x = \delta_1(\varepsilon)$ and $x = \delta_2(\varepsilon)$ in terms of an arbitrary ε .

SOME HELP IN ANSWERING PART (b)

Conceptually, this problem is very much like part (a), but it is extremely complicated to solve the two cubic equations

$$f(x) = 11 \pm \varepsilon$$

by hand and get the endpoints in terms of ε . However, Maple knows Cardan's formula for solving a cubic equation. The `fsolve` command used in part (a) solves equations numerically, but the `solve` command solves equations symbolically.

The left endpoint of the largest interval for a successful weld is found by using the Maple command

> delta1 := solve(f = 11 − eps, x);

Here we use eps to designate ε . Maple returns the three solutions of the cubic equation as a list. Notice that the first root is real and the other two roots are complex. The roots are separated by commas, and the I stands for the imaginary number i , that is, $\sqrt{-1}$. We are interested in the real root, and to obtain it we use the command

> d1 := delta1[1];

$$d1 := (5/108 - 1/4\ eps + 1/36\ (1 - 10\ eps + 27\ eps^2)^{1/2}\ 3^{1/2})^{1/3}$$
$$+ (5/108 - 1/4\ eps - 1/36\ (1 - 10\ eps + 27\ eps^2)^{1/2}\ 3^{1/2})^{1/3} + 2/3$$

which sets d1 equal to the first solution in the list.

In the following space finish writing down the Maple command that will yield delta2 as the x-coordinate of the point where f(x) = 11 + eps .

> delta2 :=

Upon entering this into Maple, you should get delta2 as a list, and the first entry delta2[1] should be the real root. As a check you should get the following expression for delta2[1] :

$$(5/108 + 1/4\ eps + 1/36\ (1 + 10\ eps + 27\ eps^2)^{1/2}\ 3^{1/2})^{1/3}$$
$$+ (5/108 + 1/4\ eps - 1/36\ (1 + 10\ eps + 27\ eps^2)^{1/2}\ 3^{1/2})^{1/3} + 2/3$$

Again, it is convenient to name this real root as d2 using the command

> d2 := delta2[1];

(c) The open interval (d1 , d2) is the largest interval about x = 1 that will yield a successful weld. In the following space write down what you think happens to the length of this interval as ε approaches 0 .

Construct a plot that gives the length of this interval for values of ε between 0 and 1 . Do not be intimidated by the rather complicated expressions for the endpoints d1 and d2 . They are just functions of eps .

SOME HELP FOR PART (c)

The length of the interval is d2 − d1, so just issue the command

> plot(d2 − d1, eps = 0..1, title = `Plot for Part (c)`);

It is natural in this problem to want to define the quantity d2 – d1 , and "length" might seem a reasonable choice. However, **do not do this!** Maple has certain variable names reserved, and "length" is one of those names.

(d) The intervals you found in parts (a) and (b) are not symmetric about the point x = 1 . For the case considered in part (b) where ε was arbitrary, find the largest interval *symmetric* about x = 1 that will yield a successful weld. Specify this symmetric interval by giving a δ depending on ε (or eps) so that if | x – 1 | < δ then the weld is successful.

A LITTLE HELP FOR PART (d)

Knowing d1 and d2 from the preceding, δ can be found by examining d2 – 1 and 1 – d1 . In the following space explain why $\delta = d2 – 1$.

(e) For a strip weld extending a distance of ε above and below the straight wire to be successful, the two wires must be within ε of one another, that is,

$$| f(x) – 11 | < \varepsilon .$$

After reading the definition of a limit in your textbook, briefly explain why the work you have done above shows that

$$\lim_{x \to 1} f(x) = 11 .$$

Laboratory Problems

The two problems here apply the ideas developed in the preceding example. Problem 1 is similar to the problem in the example, and Problem 2 extends the approach.

1. Consider the function

$$f(x) = x^3 + 2x + 9 .$$

It is not hard to see that

$$\lim_{x \to 1} f(x) = 12 .$$

(a)   How close to 1 does x have to be to guarantee that the value of $f(x)$ is within 0.25 of 12?

(b)   Given an arbitrary $\varepsilon > 0$, find a $\delta > 0$ depending on ε so that $|x - 1| < \delta$ implies $|f(x) - 12| < \varepsilon$.

2. In this problem you are asked to use a geometric argument to show that

$$\lim_{x \to 0} \frac{\sin(x)}{x} = 1 .$$

(a)   Find the largest interval about $x = 0$ so that $\sin(x)/x$ is within 0.001 of 1 .

(b)   Given an arbitrary $\varepsilon > 0$, find $\delta > 0$ so that

$$|x - 0| < \delta \text{ implies } |\frac{\sin(x)}{x} - 1| < \varepsilon .$$

Hint: What makes this problem more difficult than those considered earlier is that the two equations

$$\frac{\sin(x)}{x} = 1 \pm \varepsilon$$

cannot be solved analytically for x . For example, if you try to solve

$$\frac{\sin(x)}{x} = 1 - \text{eps}$$

using the Maple command

```
> solve( sin(x)/x = 1 – eps , x );
```

no response is given. It turns out that around $x = 0$, the function $\sin(x)/x$ looks like $1 - x^2$. To see how they compare, and to get an idea of how to find δ, try plotting both functions using the command

```
> plot( { sin(x)/x, 1 – x^2 } , x = –6..6, –0.2..1.1 );
```

You may also find it helpful to plot these functions over smaller x intervals.

(c)   Given an arbitrary $\varepsilon > 0$, find as large a δ as you can so that if x is within δ of 0, then $\sin(x)/x$ is within ε of 1 .

4

Tangent Lines and Smooth Curves

Objective

To use Maple to discover the slope of a line tangent to a curve without using derivatives.

In this Lab you will be asked to find the slope of a line tangent to a curve using only basic principles of geometry. You will investigate what it means for a curve to be "smooth" at a point and how this relates to the tangent line.

Laboratory Problems

1. Plot a graph of the function

$$f(x) = \frac{|\,2 - x\,|}{|\,x^3 - 4x^2 + 3x + 3\,|}$$

over the interval $[-3\,,4]$. Be sure to restrict the range of the vertical axis so you can get a useful plot.

Briefly explain why the function f approaches infinity as x approaches a certain point P between −1 and 0 . What is the value of P ?

2. At the point x = −2 , for example, the graph of f(x) looks "smooth." To get more detail of the graph near x = −2 , use the `plot` command to look at graphs of f in smaller and smaller intervals near x = −2 .

    Solution Guidelines: Start by plotting f over the interval [−2 , −1], and then reduce the interval by moving the right endpoint closer to −2 . More specifically, plot f over each of the 10 successively smaller intervals

    $$[-2 , -1] , \ [-2 , -2 + 1/2] , \ [-2 , -2 + 1/3] , \dots , [-2 , -2 + 1/10] .$$

    This can easily be done in Maple using the "do-loop" construction as illustrated in the following after we enter the formula for f into Maple.

    > f := abs(2 − x)/abs(x**3 − 4*x**2 + 3*x + 3);

    > # The "do-loop" that will plot each of the graphs is

    > for n from 1 to 10 do

    >   plot(f, x = −2..−2+1/n)

    > od;

    Notice that the semicolon is not used until after the last line. The way this construct works is that for each integer value of n from 1 to 10 Maple executes the commands between the "do" and the "od."

    Caution: After completing the do-loop, Maple leaves the value of n to be the next integer value not used in the loop. So in this case Maple leaves n = 11 . To check this, we ask Maple what value it has for n by issuing the command

    > n;

    > *11*

    If we use n in further work, n will equal 11, so we reset n to be a variable by issuing the command

    > n := 'n';

    > *n := n*

3. Notice that as the interval gets smaller the graphs in Problem 2 look more and more like the graph of a straight line, although they are slightly curved.

From the last graph over the interval $[-2, -1.9]$ get an estimate of the rise and calculate

$$\text{slope} = \frac{\text{rise}}{\text{run}} = \frac{f(-2 + 1/10) - f(-2)}{1/10}.$$

You should get, approximately,

$$\text{slope} = (.1624 - .1482)/.1 = .142.$$

Find a more precise value for the rise over the interval $[-2, -1.9]$ by using the `subs` command to evaluate f at $x = -2$ and $x = -1.9$, and recalculate a more precise value of the slope. You should get

$$\text{slope} = .143586230.$$

4. Write down the equation of the straight line that passes through the point $(-2, f(-2))$ and has slope $m = 0.143586230$. Recall that if a line passes through the point (a, b) and has slope m, then

$$\frac{y - b}{x - a} = m, \quad \text{so} \quad y = b + m*(x - a) \text{ is its equation.}$$

Plot this line and the graph of f on the same axes over the interval $[-3, -1]$. Your line should appear to be a good approximation to the curve for values of x near $x = -2$.

5. By construction, the line we produced in Problems 3 and 4 passes through the points $(-2, f(-2))$ and $(-1.9, f(-1.9))$. Briefly explain why this is the case. However, the line is not tangent to the graph of f at $x = -2$. Construct a plot that shows this.

6. We know that the line tangent to the graph at the point $(-2, f(-2))$ goes through the point $x = -2$ and $y = f(-2)$. So the reason that the line we constructed is not actually tangent to the graph of f is that we must have used the wrong value for the slope m.

   How could you obtain a more precise value for the true slope of the line tangent to the graph of f at $x = -2$? Find a better estimate, and find the resulting line. Then, on the same graph, plot the line, the line from Problem 4, and the function f. Your plot should show that your new line is more nearly tangent to the graph than the line constructed in Problem 4.

7. How would you use a limiting process to find the exact slope of the line tangent to the graph of f at $x = -2$? Recall that before we used the estimate

$$\text{slope} \;=\; \frac{f(-2 + 1/n) - f(-2)}{1/n} \quad\quad \text{for } n = 10 \;.$$

Use the `limit` command in Maple to find the slope of the line tangent to the graph at $x = -2$.

8. The graph of a function is called "smooth" at a point if the graph has a unique tangent line at the point. The graph of f is surely smooth at $x = -2$. If a graph is smooth at a point, then as we look at the graph in smaller and smaller intervals about the point, the graphs look more and more like straight lines whose slopes get closer and closer to the slope of the line tangent to the graph at that point. Explain why the graph of f is smooth at $x = 3$, and give an estimate of the slope of the line tangent to the graph at the point $(3 , f(3))$. Is the graph of f smooth at $x = 2$? Explain.

   Is the graph of f smooth at point P in Problem 1 where f approaches $+\infty$? Explain.

5

Derivatives and Tangent Lines

Objective

To use Maple to solve some typical problems involving differentiation and the construction of tangent line approximations.

This Laboratory begins with two examples that involve differentiation, solving equations, and graphing. Following the statement of these examples are their detailed solutions. You should work through these examples and their solutions as a means of learning more about Maple. This will help you in doing the Laboratory problems.

EXAMPLE 1

Let $f(x) = 4x^3 - 4x^2 - 11x + 7$.

    (a) Find the roots of $f(x) = 0$.

    (b) Plot the graph of $y = f(x)$ on an x interval containing all of the roots.

    (c) Plot the graph of $y = f(x)$ on $x = -2..0$. Then plot several more graphs, using shorter and shorter intervals all of which are centered at $x = -1$. Describe the behavior of the function f that you see depicted in these graphs.

(d) Find an equation of the line tangent to the graph of $y = f(x)$ at $x = -1$. Plot the graphs of this line and of $y = f(x)$ on a short interval surrounding $x = -1$.

EXAMPLE 2

Let $f(x) = 2x^2 + x - 4$, $g(x) = \sin(2x)$, and $h(x) = \text{sqrt}(4 - 3x^2)$. Use Maple to calculate each of the following derivatives:

(a) $\dfrac{d}{dx}\,(f(x)g(x))$,   (b) $\dfrac{d}{dx}\,\dfrac{f(x)}{g(x)}$,   (c) $\dfrac{d^2}{dx^2}\,\dfrac{f(x)}{g(x)}$,

(d) $\dfrac{d}{dx}\,\dfrac{g(x)}{f(x)h(x)}$,   (e) $\dfrac{d}{dx}\,f(g(x))$,   (f) $\dfrac{d}{dx}\,f(g(h(x)))$.

Solution of Example 1   First we enter the formula for f:

> f := 4*x^3 – 4*x^2 – 11*x + 7;

$$f := 4\,x^3 - 4\,x^2 - 11\,x + 7$$

(a) There are two commands that can be used to find the roots. One is `fsolve`:

> fsolve(f, x);

$$-1.534758996,\ .5847327224,\ 1.950026274$$

Note that `fsolve` gives the roots as decimal numbers. The other command is `solve,` and this will find the roots exactly since f is a cubic. To demonstrate:

> solve(f, x);

$$\%2 + \%1 + 1/3,\ -1/2\,\%2 - 1/2\,\%1 + 1/3 + 1/2\,3^{1/2}\,(\%2 - \%1)\,I,$$

$$-1/2\,\%2 - 1/2\,\%1 + 1/3 - 1/2\,3^{1/2}\,(\%2 - \%1)\,I$$

$$\%1 := \left(-\frac{41}{108} - 1/72\,I\,1627^{1/2}\,3^{1/2}\right)^{1/3}$$

$$\%2 := \left(-\frac{41}{108} + 1/72\,I\,1627^{1/2}\,3^{1/2}\right)^{1/3}$$

Which would you rather deal with, the exact formulas or the numerical approximations? In the expressions given by `solve`, the three roots are separated by commas, and the quantities denoted by %1 and %2, respectively, are listed next. Observe that the expressions for two of the roots contain I , the imaginary unit, although the roots are actually real numbers (as `fsolve` shows).

(b) A convenient interval containing the roots is $x = -2..3$.

> plot(f, x = −2..3, title = `A Plot of f := 4*x^3 − 4*x^2 − 11*x + 7`);

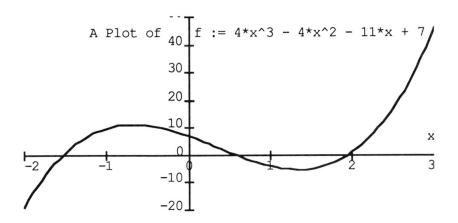

To the right of $x = 3$ the graph of $y = f(x)$ continues to rise steeply, and similarly to the left of $x = -2$. Thus the interesting behavior of the graph is confined to an interval such as the one used here.

(c) First we plot the function over $-2 \le x \le 0$:

> plot(f, x = −2..0);

Now let us successively shorten the interval by approximately a factor of two:

> plot(f, x = −1.5..−0.5);
> plot(f, x = −1.2..−0.8);
> plot(f, x = −1.1..−0.9);
> plot(f, x = −1.05..−0.95);
> plot(f, x = −1.02..−0.98);

These plots show that the graph of $y = f(x)$ becomes more and more like a straight line as the interval is reduced in length.

(d) The equation of the tangent line at a point x0 is y − y0 = m(x − x0) ,
where y0 = f(x0) is the value of the function f at x0 , and m = f '(x0) is the slope
at that point. Now we calculate these quantities for the given function f :

> x0 := −1; y0 := subs(x = x0, f);

$$x0 := -1$$
$$y0 := 10$$

To find the slope we must differentiate f :

> fprime := diff(f, x);

$$fprime := 12\,x^2 - 8\,x - 11$$

> m := subs(x = x0, fprime);

$$m := 9$$

Thus, the equation of the tangent line is y = g(x) , where g(x) = y0 + m(x − x0) :

> g := y0 + m*(x − x0);

$$g := 19 + 9\,x$$

Now we plot both y = f(x) and y = g(x) on a short interval about x0 = −1 . To
plot two graphs simultaneously we must enclose the two expressions in braces, as in
the following command:

> plot({f, g}, x = −1.05..−0.95, title = `A Function and Its Tangent Line`);

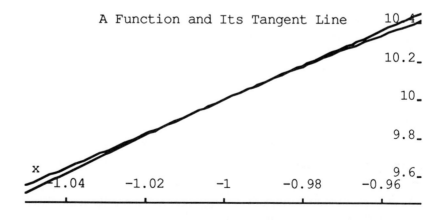

The tangent line is a good approximation to $y = f(x)$ on this interval. On a somewhat shorter interval the two graphs would be indistinguishable to within the resolution of the screen. This property of $y = f(x)$ near $x = -1$ is summarized by saying that f is **locally linear** there. An important mathematical fact is that every differentiable function is locally linear in the neighborhood of any point where the derivative exists.

Solution of Example 2   First we enter expressions for $f(x)$, $g(x)$, and $h(x)$:

> f := 2*x^2 + x − 4; g := sin(2*x); h := sqrt(3 − 4*x^2);

$$f := 2\,x^2 + x - 4$$

$$g := sin(2\,x)$$

$$h := (3 - 4\,x^2)^{1/2}$$

(a) It is very easy to calculate the derivative using the `diff` command:

> diff(f*g, x);

$$(4\,x + 1)\,sin(2\,x) + 2\,(2\,x^2 + x - 4)\,cos(2\,x)$$

Clearly Maple knows the product rule for derivatives.

(b) From this point on we will not include Maple's response, but you can find out what it is by just entering the command

> deriv := diff(f/g, x);

It appears that Maple has also differentiated this quotient using the product rule. We can rewrite the result as follows:

> normal(deriv);

Does this look more like the result of using the quotient rule?

(c) One can either find the second derivative by differentiating the first derivative, or by asking Maple to find the second derivative directly. We will show both ways here.

> deriv2 := diff(deriv, x);

To have Maple take the second derivative directly we include x$2 in the `diff `
command. The $2 instructs Maple to repeat the operation (i.e., it is done twice).
In a similar way $n would cause it to be done n times.

> diff(f/g, x$2);

By comparing terms you can see that the two expressions for the second derivative
are the same.

(d) In the command that follows note two things. First, parentheses are needed
around f * h in the denominator. Second, we will suppress the response (to save
space) by ending the command with a colon. If you want to see the result, you
should replace the colon by a semicolon.

> diff(g /(f *h), x):

(e) This is an illustration of the chain rule. It is helpful to redefine f , g , and
h as functions, rather than as expressions:

> f := x –> 2*x^2 + x – 4; g := x –> sin(2*x); h := x –> sqrt(3 – 4*x^2):

Then,

> diff(f(g(x)), x):

(f) Another use of the chain rule, this time applied twice:

> diff(f(g(h(x))) , x):

Note: Differentiation is a process that, in most instances, involves only a few rules,
which are used over and over. In addition, even for relatively simple functions,
such as those used in this example, the results quickly become rather complicated
and unwieldy. Therefore differentiation lends itself very well to execution by a
computer. Although you should know the basic rules of differentiation and how to
apply them, you should also think of Maple when the calculations begin to become
complicated.

Laboratory Problems

These problems apply the ideas discussed in the preceding examples. They are
independent of one another, although Problem 2 refers to instructions given in
Problem 1.

1. Let $f(x) = 4x^4 + 4x^3 - 13x^2 - 7x + 8$.

   (a) Find the roots of $f(x) = 0$.

   (b) Plot the graph of $y = f(x)$ on an x interval that includes all of the roots.

   (c) Consider values of x near $x = -1/2$. Plot the graph of $y = f(x)$ on an interval of length 2 centered at $x = -1/2$. Then plot several more graphs of $y = f(x)$, on shorter and shorter intervals centered at $x = -1/2$. How does the graph of $y = f(x)$ behave as the interval is reduced in length? Do you think that f is locally linear in a neighborhood of $x = -1/2$?

   (d) Find an equation of the line tangent to the graph of $y = f(x)$ at x = $-1/2$. Plot the graphs of the tangent line and of $y = f(x)$ on a short interval about $x = -1/2$. On what interval are the graphs virtually indistinguishable?

2. Let $f(x) = (\sin(x^2))/x$.

   (a) Find the first two positive roots of $f(x) = 0$.

   (b) Plot the graph of $y = f(x)$ on an interval containing the roots found in part (a).

   (c) Consider values of x near $x = 1$, and follow the instructions in part (c) of Problem 1.

   (d) Follow the instructions of part (d) of Problem 1, using the point $x = 1$.

3. Choose a differentiable function of your own. Select a point and try to determine whether your function is locally linear near that point. Over what interval is the tangent line at that point a good approximation to your function?

4. Let $f(x) = \cos(x/2)$, $g(x) = x^2 - 3x + 1$, and $h(x) = \text{sqrt}(x^2 + 2x)$. Use Maple to calculate each of the following derivatives:

   (a) $\dfrac{d}{dx} (f(x)g(x))$,   (b) $\dfrac{d^2}{dx^2} (f(x)g(x))$,   (c) $\dfrac{d}{dx} \dfrac{f(x)g(x)}{h(x)}$,

   (d) $\dfrac{d}{dx} f(g(x))$,   (e) $\dfrac{d^2}{dx^2} f(g(x))$,   (f) $\dfrac{d}{dx} f(g(h(x)))$.

6

Introduction to Maxima and Minima

Objective

To use Maple to find the maxima and minima of functions and to use this information in constructing plots.

The purpose of this Laboratory session is to begin exploring Maple's capabilities for studying maxima and minima of functions. These are needed in many mathematical problems and applications (e.g., see Lab 7) and may also be helpful in sketching a useful graph of a function. The interplay between the graph or geometric properties of functions and the algebraic and analytic operations performed on them is one of Maple's most important capabilities.

Following you will find two examples that are worked out in detail, and these are followed by the Laboratory problems.

EXAMPLE 1

Let $f(x) = x(2x - 3)(x + 3.8)$.

    (a) Draw the graph of this function.

    (b) Find the local maxima and minima of f.

    (c) Draw the graphs of f and f' on the same axes. How are the two graphs related?

(d) Also draw the graphs of f and f″ on the same axes. How are these graphs related? Discuss the concavity of f and find any inflection points of the graph of f.

EXAMPLE 2

Let $h(x) = x^5 - x^4 - 15x^3 - x^2 + 28x + 7$.

(a) Find the roots of $h(x) = 0$.

(b) Plot the graph of $y = h(x)$.

(c) Find $h'(x)$ and plot the graph of $y = h'(x)$.

Solution to Example 1   First we define the function f.

> f := x*(2*x − 3)*(x + 3.8);

$$f := x\,(2\,x - 3)\,(x + 3.8)$$

Let us reduce the number of digits to six, to cut down the amount of computing time. This is done with the command

> Digits := 6;

$$Digits := 6$$

(a) Before commanding Maple to plot a graph of f, let us try to sketch the graph using information about the function. (The exercise of trying to visualize the graphs of functions is useful in enhancing your understanding of functions and their properties.)

Draw a rough sketch of the graph of f, making use of these facts: (i) The graph crosses the x-axis at the zeros of f(x), x = −3.8, 0, 3/2, and only at these points (how do we know this?); (ii) The graph is a smooth, continuous curve (because f is a polynomial); and (iii) As x gets very large and positive (negative), so does f(x). (Why?)

Now let Maple plot the graph of f.

> plot(f);

When the `plot` command is given without specification of an x interval (as was just done), Maple always uses the x interval [−10, 10], the so-called "default" interval. This graph is not useful because the scale is so compressed that interesting features of the graph are not shown. The situation can be improved by using a shorter x interval, but one that includes all of the roots. Let us try the interval −4..3 (i.e., $-4 \le x \le 3$).

> plot(f, x = −4..3, title = `Plot for Example 1(a)`);

Does your hand drawn graph bear a reasonable resemblance to the last graph produced by Maple?

(b) We wish to find the local extreme points of f . First we shall do this approximately by examining the last Maple plot (i.e., the graph of f on the interval −4 ≤ x ≤ 3).

You should see two local extrema on this graph. Do you think that there may be more, which don't appear here? Explain your answer in the space here:

What is the largest number (and the smallest number) of local extreme points that the graph of a cubic polynomial can have? Explain your answer in the following space:

From this Maple graph, estimate the coordinates of each local maximum or minimum, and enter the coordinates of each point (labeled "max" or "min") in the following space:

A better plot of f may be obtained by adjusting the y interval. Before doing this it is useful to take account of the y values at the extreme points. From the Maple plot the local maximum appears to be somewhat less than 30 and the local minimum a little above −10 . This leads to the following plot:

> plot(f, x = −4..3, y = −10..30, title = `Plot for Example 1(b)`);

To locate the extreme points more precisely, we call on Maple to find f′(x) and then to solve f′(x) = 0 :

> fprime := diff(f, x);

$$fprime := (2\,x - 3)\,(x + 3.8) + 2\,x\,(x + 3.8) + x\,(2\,x - 3)$$

Is Maple's result correct? Check it by hand.

> roots := solve(fprime, x);

$$roots := .810603, -2.34394$$

These are the two critical points where $f'(x) = 0$. Let's call them x1 and x2 for reference, x1 being the smaller. Since this root is listed second, we designate it by 'roots[2]' .

> x1 := roots[2];

$$x1 := -2.34394$$

> x2 := roots[1];

$$x2 := .810603$$

The corresponding y values can now be found:

> y1 := subs (x = x1, f);

$$y1 := 26.2381$$

> y2 := subs (x = x2, f);

$$y2 := -5.15304$$

How do the coordinates of the local maximum (x1 , y1) and the local minimum (x2 , y2) compare with your estimates?

   (c) Now let's plot the graphs of f and fprime on the same axes. It turns out that the y interval of $-10..30$ cuts off part of the graph of fprime . By proceeding as before to study the function fprime , or by trial and error, one can readily find an interval that gives a good plot; for example, the y interval of $-20..30$ is adequate.

> plot({f, fprime}, x = $-4..3$, y = $-20..30$, title = `Plot for Example 1(c)`);

Observe that when fprime < 0 (so its graph is below the x-axis), which occurs when x is between x1 and x2 , then the graph of f is decreasing. On the other hand, when fprime > 0 (so its graph is above the x-axis), then the graph of f is increasing. Compare the behavior of the graph of fprime in the neighborhood of the maximum point x1 with its behavior in the neighborhood of the minimum point x2 .

   (d) Now let us find the second derivative of f . First do it with pencil and paper and write the result in the following space:

Now let Maple do it:

> f2prime := diff(fprime, x);

$$f2prime := 12. \, x + 9.20000$$

Does your expression for $f''(x)$ agree with Maple's result?

Next we plot the graphs of f and f2prime on the same axes:

> plot({f, f2prime}, x = −4..3, y = −20..30, title = `Plot for Example 1(d)`);

Observe that the graph of f is concave up (like a cup) when f2prime is positive, and that the graph f is concave down (like a cap) when f2prime is negative. Also notice that f2prime is positive at the minimum point and that f2prime is negative at the maximum point.

To find the point where the graph of f changes its concavity (an inflection point) we must find where f2prime is zero:

> x3 := solve (f2prime, x);

$$x3 := -.766667$$

> y3 := subs (x = x3, f);

$$y3 := 10.5425$$

Solution to Example 2   This example involves a fifth-degree polynomial. Now we shall begin to appreciate the power of Maple to deal with fairly complicated functions.

> h := x ∧ 5 − x ∧ 4 − 15*x ∧ 3 − x ∧ 2 + 28*x + 7;

$$h := x^5 - x^4 - 15x^3 - x^2 + 28\,x + 7$$

(a) Can you find the roots of h(x) = 0 by pencil and paper? Not likely. Cubic and higher degree polynomials are difficult to solve by hand except in special cases.

We ask Maple to find the roots by using the `solve` command:

> solve(h, x);

$$RootOf(\_Z^5 - \_Z^4 - 15\_Z^3 - \_Z^2 + 28\_Z + 7)$$

This is not very helpful, since it is just a restatement of the original problem, using _Z in place of x . What has happened is that the `solve` command tells Maple to find the roots exactly, and there is no general way to do this for a fifth-degree polynomial.

There is another command, `fsolve,` that tells Maple to find numerical approximations to the real roots. This is the command that we need here:

> fsolve(h, x);

$$-2.96299, -1.47680, -.256496, 1.47881, 4.21748$$

Maple has found five roots, each calculated to six digits, the current value of Digits. Now you can see why we did not ask you to find the roots of $h(x) = 0$ by hand.

(b) Next we want to plot the graph of h. Using the roots of $h(x) = 0$ we have listed and the behavior of $h(x)$ as $|x|$ gets large, make a rough sketch of the graph of h. Note that we cannot tell how high or low the graph gets in various intervals without further work.
Let us now use Maple to plot the graph of h. We choose an interval that includes all of the roots:

> plot(h, x = −3..5, title = `Plot for Example 2(b)`);

The interesting features of the graph are rather compressed; we can display them more clearly by specifying an interval for y. By trial and error we find that an interval that includes the local maxima and minima of the function is $-225..50$:

> plot(h, x = −3..5, y = −225..50, title = `Another Plot for Example 2(b)`);

How does this plot compare with your sketch?

(c) Find $h'(x)$, the derivative of $h(x)$, and write the result in the following space:

Finally, we ask Maple to find $h'(x)$ and then plot its graph.

> hprime := diff(h, x);

$$hprime := 5\,x^4 - 4\,x^3 - 45\,x^2 - 2x + 28$$

Does your result for $h'(x)$ agree with Maple's?

> plot(hprime, x = −3..5, title = `Plot for Example 2(c)`);

> quit

Laboratory Problems

In this Lab you are to carry out steps similar to those in the examples. The three problems are independent of one another.

1. Let $f(x) = 2x^4 - 4x^3 - 11x^2 + 8x + 4$.

    (a) Draw the graph of $y = f(x)$ for $-2 \le x \le 3.5$.

    (b) Find the roots of $f(x) = 0$.

    (c) Find $f'(x)$.

    (d) Draw the graphs of $f(x)$ and $f'(x)$ on the same set of axes.

    (e) Find the maximum and minimum points of $f(x)$.

2. Let $f(x) = x^5 - 5x^4 + 5x^3 + 7$.

    (a) Plot the graph of f using the default interval $-10..10$. (Thus you have only to use the command `plot(f)`.)

    The next two parts produce information useful in constructing a better graph.

    (b) Find the roots of $f(x) = 0$.

    (c) Find the critical points, that is, solve $f'(x) = 0$, and determine the local maximum and minimum values of f.

    (d) Based on the results of (b) and (c) choose x and y intervals that lead to a graph of f that clearly shows its major features and plot the graph.

3. Let $f(x) = (x^2 - 2x)/(x^2 + 4)$.

    (a) Plot the graph of f for $x = -10..10$.

    (b) Estimate the location of all maximum, minimum, and inflection points. How does the graph behave for large positive x? For large negative x?

    (c) Use Maple to determine the maximum, minimum, and inflection points.

7

Applications of Max-Min Problems

Objective

To use Maple to solve some typical max-min problems.

This Laboratory extends our investigation of max-min problems, in two examples and three problems. The reason we devote so much attention in the calculus to max-min problems is that they are examples of what are called "optimization" problems, which are of fundamental importance in economics, science, and engineering. The task in an optimization problem is to obtain the best possible result with limited resources.

It is recommended that you work through the examples, as they will prepare you for the Laboratory problems:

EXAMPLE 1

Let $f(x) = \sin x + x \cos(x^2)$ for $-\pi \le x \le \pi$.

(a) Plot the graph of f on $-\pi \le x \le \pi$, and estimate where the local maxima and minima are located.

(b) Find the local maxima and minima.

(c) Find the absolute maximum and minimum of f on the given interval.

EXAMPLE 2

Find the rectangle of largest area (located entirely in the first quadrant with sides parallel to the axes) that can be fitted under the arc of the ellipse

$$\frac{x^2}{9} + \frac{y^2}{4} = 1 .$$

Solution to Example 1   We first enter the function

> f := sin(x) + x*cos(x^2);

$$f := sin(x) + x \, cos(x^2)$$

(a) To plot the graph of f on −Pi..Pi (i.e., on $-\pi \le x \le \pi$) we use the command

> plot(f, −Pi..Pi, title = `Plot for Example 1 (a) `);

What kind of symmetry does this graph have? Could you have predicted this before seeing the graph?

(b) The function f has several local maxima and minima. You can estimate the coordinates of these points from the graph. In some versions of Maple (such as for X-windows) this can be done very easily using the mouse. For other versions you may have to do this by eye. However you do it, record the results in the following space:

To determine the maxima and minima of f more accurately, we look for points where $f'(x) = 0$. Using pencil and paper find $f'(x)$, and write the result in the following space:

Now let Maple do it:

> fprime := diff(f, x);

Does Maple's result agree with yours?

> Digits := 6;

$$Digits := 6$$

> x1 := fsolve(fprime, x);

$$x1 := .920110$$

This is one of the roots of $f'(x) = 0$. (In general, in solving a non-polynomial equation, Maple can find only one root at a time.) Next we find the corresponding value of y :

> y1 := evalf(subs(x = x1, f));

$$y1 := 1.40527$$

Is the point $(x1, y1)$, just determined by use of Maple, close to one of those you found above by examination of the graph?

From the graph it is clear that $(x1, y1)$ is a local maximum. By symmetry the point $(-x1, -y1)$ is a local minimum. To find another critical point we must use `fsolve` again, specifying an interval that excludes the root already found (that is, x1) and including at least one other root. The estimates you made earlier should help you to do this. For example, the following command works:

> x2 := fsolve(fprime, x, 1..2);

$$x2 := 1.82428$$

> y2 := evalf(subs(x = x2, f));

$$y2 := -.824635$$

Now we repeat the process to find a third critical point:

> x3 := fsolve(fprime, x, 2..3);

$$x3 := 2.50968$$

> y3 := evalf(subs(x = x3, f));

$$y3 := 3.10008$$

The graph indicates that there is another critical point in the open interval 2.7..Pi .

> x4 := fsolve(fprime, x, 2.7..Pi);

$$x4 := 3.08700$$

> y4 := evalf(subs(x = x4, f));

$$y4 := -3.01549$$

Finally we evaluate f at the endpoint:

> y5 := evalf(subs(x = Pi, f));

$$y5 := -2.83589$$

(c) Taking advantage of the symmetry of the function, we conclude that the absolute maximum on −Pi..Pi is at the point x3 , where f has the value 3.10008 . Similarly, the absolute minimum is at the point −x3 , where f has the value −3.10008 .

Before leaving this function, let us plot the graph of f′(x) :

> plot(fprime, x = −Pi..Pi, title = `Plot for Example 1 (c) `);

Observe that this graph also has a symmetry property; specifically the graph of f′(x) is symmetric about the y-axis (that is, f′(x) is an even function), whereas the graph of f is symmetric about the origin (that is, f is an odd function). Is this accidental? Or is the derivative of an odd function always even? Can you show this?

What about the derivative of an even function? Since f′(x) is an even function, you might take its derivative and plot its graph to see if the result supports your answer.

Solution to Example 2   Here a strategy, or plan of action, is needed. We want to know, in effect, which value of x in 0 < x < 3 gives the rectangle of largest area under the ellipse, as shown in the following figure. For each x there is a rectangle whose base and altitude (therefore, whose area) depend on x . When the area is expressed in terms of x , it is a function of x , which we then seek to maximize by standard methods (setting its derivative equal to zero, etc.).

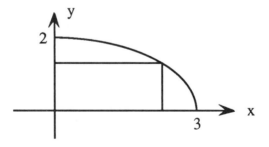

To get an expression for the area of the rectangle (of base x) with one vertex on the ellipse, we need an expression for the height of the rectangle, obtained by solving the equation of the ellipse for y in terms of x . This is how we shall start.

Note that our strategy has been developed by working backwards: focusing initially on the objective and asking what prior steps have to be carried out in order to reach the objective.

Let us assign the name 'eq' to the given equation (of the ellipse):

```
> eq := (x^2)/ 9 + (y^2)/4 = 1;
```
$$eq := 1/9\, x^2 + 1/4\, y^2 = 1$$

Solve this equation for y as a function of x and write the result in the following space:

Now we ask Maple to solve the equation for y . (*Note:* Maple cannot plot equations given in the form F(x , y) = 0 .)

```
> solve(eq, y);
```

Observe that Maple has found two solutions. Do your results agree with Maple's?

Of the two expressions for y given by Maple, we want the one for which y > 0 . We call this function f :

```
> f := "[1];
```
$$f := 2/3\, (-x^2 + 9)^{1/2}$$

We are now at the crucial part of the problem: setting up the function that is to be maximized. Since the rectangle of base x has height f , its area is the product of f and x . Thus the function to be maximized is x * f . We call this function g :

```
> g := x * f;
```
$$g := 2/3\, x\, (-x^2 + 9)^{1/2}$$

To study the behavior of g , let us graph it. The appropriate interval for x is 0..3 . (Why?)

```
> plot(g, x = 0..3, title = `Plot for Example 2`);
```

Clearly g has a single maximum in the vicinity of x = 2 . Estimate its location from the graph. To find it accurately, we take the derivative of g with respect to x:

```
> gprime := diff(g, x);
```

Next, we solve the equation gprime = 0 . (Could you solve this by hand?)

```
> xzeros := solve(gprime, x);
```

$$xzeros := -1/2\ 9^{1/2}2^{1/2}\ ,\ 1/2\ 9^{1/2}2^{1/2}\ ,\ -1/2\ 9^{1/2}2^{1/2}\ ,\ 1/2\ 9^{1/2}2^{1/2}$$

There are only two distinct solutions of the equation, and we are interested in the positive one. We will designate this solution $xcrit$, and so

> xcrit := "[2];

$$xcrit := 1/2\ 9^{1/2}2^{1/2}$$

> evalf(");

$$2.12132$$

The area of the rectangle corresponding to $xcrit$ (that is, with base from 0 to xcrit) is the value of g at $x = xcrit$. We call this value A:

> A := subs(x = xcrit, g);

$$A := 3$$

From the graph it appears that this is a maximum rather than a minimum. Let us check this conclusion analytically, in two ways. First we use the second derivative test:

> g2prime := diff(gprime, x);

$$g2prime := -\ 2\ \frac{x}{(-x^2+9)^{1/2}}\ -\ 2/3\ \frac{x^3}{(-x^2+9)^{3/2}}$$

> subs(x=xcrit, g2prime);

$$-8/3$$

A negative value for the second derivative signifies a maximum.

As an alternative method of checking the observation that g has a local maximum at $x = xcrit = 2.12132$, let us use the idea, though not the precise statement, of the first derivative test. Specifically, let us evaluate $gprime$ at two values of x near $x = xcrit$, one to the left of $x = xcrit$, the other to the right. We choose $x = 2$ as the point to the left, and $x = 2.25$ as the point to the right, of $x = xcrit$:

> evalf(subs(x = 2, gprime));

$$.298142$$

> evalf(subs(x = 2.25, gprime));

$$-.37798$$

The first number is positive, the second one is negative, which should be the case if g is maximized at $x = xcrit$.

Let us now find the y value corresponding to $xcrit$:

> ycrit := subs(x = xcrit, f);

$$ycrit := 1/3 \ 9^{1/2} \ 2^{1/2}$$

Finally we calculate the ratio of xcrit to ycrit :

> xcrit/ycrit;

$$3/2$$

Observe that this value is the same as the ratio of the two semiaxes of the ellipse. Is this a coincidence?

> quit

Problem 3 below involves extensions of the work done on this example.

Laboratory Problems

In this Lab you are to carry out steps similar to those in the example. The three problems are independent of one another.

1. Let $f(x) = x^2 e^{-2x}$.

   (a) Find the critical points and the local maxima and minima of f .

   (b) Using the information from part (a), draw the graph of f .

   (c) Plot the graphs of f and f' on the same axes.

   (d) Test the critical points analytically by evaluating f''(x) at each one.

2. The following curves connect the two points $(0, 2)$ and $(3, 0)$ in the first quadrant:

$$\text{C1: } y = 2 \ \sqrt{1 - (x/3)},$$

$$\text{C2: } y = 2 \cos(\pi x/6) .$$

   We wish to fit rectangles under each of the curves; the rectangles are to have two sides along the axes with one corner at the origin.

   (a) Plot the two curves on the same axes. Can you tell which curve can have the larger rectangle fitted under it?

(b) Find the area of the largest rectangle that can be fitted under each curve.

3. In Example 2 we considered the ellipse $x^2/9 + y^2/4 = 1$ and the rectangle of largest area that could fit under it in the first quadrant with its sides parallel to the coordinate axes. We wish to extend some of that work to an arbitrary ellipse, written in the standard form

$$\frac{x^2}{a^2} + \frac{y^2}{b^2} = 1 , \qquad (1)$$

where $a > 0$ and $b > 0$ are constants.

(a) Here we shall be interested in ratios of areas; in this connection it should be noted that the area enclosed by the ellipse described by Eq. (1) is $ab\pi$. Thus for the ellipse studied in Example 2 (where $a = 3$, $b = 2$) the area in the first quadrant enclosed by the ellipse is $3\pi/2$. Since we saw that the largest inscribed rectangle has area 3, the ratio of the rectangular area to the elliptically enclosed area is $2/\pi$.

How does this ratio change, or does it remain the same, as the ellipse is varied (that is, as the parameters a and b are varied)?

(b) In Example 2 the ratio of $x$crit to $y$crit, that is, the ratio of the sides of the maximum inscribed rectangle, is the same as the ratio of the semiaxes of the ellipse, namely, $3/2$. Is this true for all ellipses? In particular, for the ellipse (1) is the ratio of $x$crit to $y$crit equal to a/b? Try to prove your answer.

8

Some Interesting Curves
Described by Parametric
Equations

Objective
To introduce parametric equations and to illustrate their use in describing some
rather intricate curves.

Frequently, a plane curve is the graph of a function $y = f(x)$ or it is determined
by an equation of the form $F(x, y) = 0$. A third way to describe a curve is by
means of a set of parametric equations, in which both coordinates x and y are
expressed as functions of a parameter. The subject of parametrized curves has
always been a part of calculus but with the introduction of computer graphics it has
become essential. It is with this in mind that one of the primary objectives of this
Laboratory is to help you develop skills for using, and understanding, parametric
equations.

An example of a parametric representation of a curve is

$$x = 3 + \sin(t), \quad y = 2 + \frac{1}{2}\cos(t), \tag{1}$$

where t is the parameter and is restricted to the interval $t = 0..2*Pi$ (i.e., $0 \le t \le 2\pi$). Note that an interval is given for t. This is a necessary part of the description
of the curve—a different interval corresponds to a different curve.

EXAMPLE 1

Suppose we want to plot the curve given in (1) for $0 \le t \le 2\pi$. To do this Maple first computes (x , y) pairs for many different values of t , it then locates the corresponding points in the x,y-plane, and finally it draws a smooth curve through them. For example, the (x , y) pair for t = 0 is (3 , 5/2) , so that is the starting point of the curve.

To use the `plot` command to draw the curve, one must simply enclose the plot information in square brackets to signify that the graph is to be drawn parametrically. Thus, the command is

> plot([3 + sin(t) , 2 + 1/2*cos(t) , t = 0..2*Pi] , title = `Figure 1`);

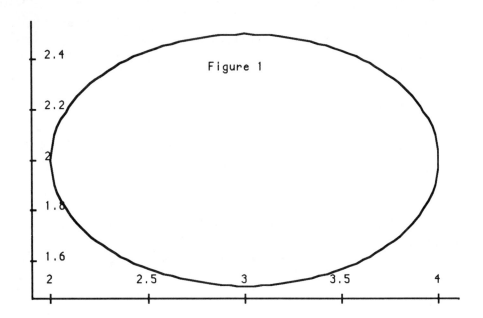

This produces the graph shown in Fig. 1. The shape of the graph is somewhat distorted since the scales on the x-and y-axes are different. One way to reduce this distortion is to specify equal lengths on both axes. Then, after the plot is displayed, we can resize the widow so it is square.

> plot([3 + sin(t), 2 + 1/2*cos(t), t = 0..2*Pi], x = 1..5, y = 0..4, title = `Figure 2`);

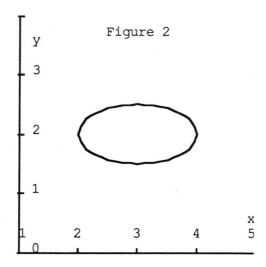

Figure 2

Figure 2 shows the actual shape of the ellipse more clearly.

Can you determine whether the curve is traversed clockwise or counterclockwise as t increases?

EXAMPLE 2

If a circle of radius a rolls without slipping along a straight line, any given point on the circumference of the circle traces a curve known as a cycloid. If the given point is initially at the origin, then the cycloid can be described by the following set of parametric equations:

$$x = a*(t - \sin(t)) , \quad y = a*(1 - \cos(t)) . \tag{2}$$

In this case the variable t is the angle between the radius to the given point and the vertical (y) axis. It is recommended that you review the discussion of parametric equations in your calculus text (looking in the index, if necessary, to find this discussion).

Suppose we want to plot the graph of Eqs. (2) for a = 1 and t = 0..4*Pi , that is, for two full revolutions of the circle. It is possible to proceed exactly as in Example 1, but, for variety, we will give a sequence of commands instead.

```
> f := a*(t - sin(t));  g := a*(1 - cos(t));
```

$$f := a\,(t - sin(t))$$
$$g := a\,(1 - cos(t))$$

```
> a := 1;
```

$$a := 1$$

```
> plot( [f, g, t = 0..4*Pi], title = `Figure 3`);
```

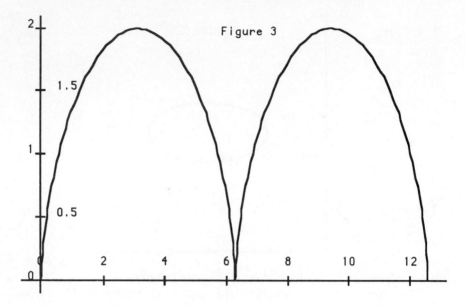

Figure 3 shows the graph of two arches of a cycloid. If a longer t interval is used, then there are more arches. Notice the sharp points (cusps) each time the graph touches the x-axis. Notice also the scales on the two axes.

Can you use the geometrical interpretation of the cycloid in terms of a rolling circle to determine (without any calculation) the maximum points on the graph?

Now, suppose we use a different value of a , such as a = 2 , and replot the graph:

> a := 2;

$$a := 2$$

> plot([f, g, t = 0..4*Pi], title = `Figure 4`);

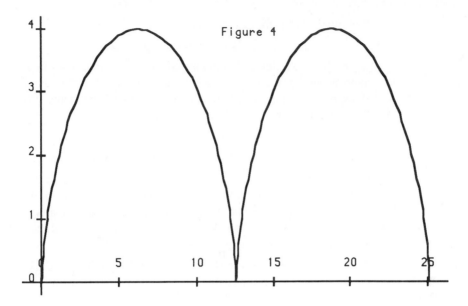

Figure 4

How does the graph in Fig. 4 compare with the one in Fig. 3? Are the scales on the axes the same?

Important conclusion: Since the constant a appears only as a multiplier in the expressions for x and y , it affects only the **size** of the graph. The **shape** of the graph is the same for all positive values of a .

Laboratory Problems

In this Lab you are to investigate parametrizations for various curves. Problems 1 and 2 deal with different curves and are therefore independent of one another.

1. Generalize the construction of the cycloid in the following way. Let the given point be at a distance b from the center of the rolling circle of radius a . If b < a , think of the point as located on a spoke of a bicycle wheel, while if b > a , imagine that the spoke is extended beyond the circumference of the circle. The curve traced out by the point as the circle rolls along the x-axis is called a *trochoid*.

    (a) Show that parametric equations of the trochoid are

    $$x = at - b \sin t , \quad y = a - b \cos t .$$ (3)

Rewrite these equations in the form

$$x = a(t - \beta \sin t), \quad y = a(1 - \beta \cos t), \tag{4}$$

where $\beta = b/a$. Since a again appears only as a multiplier in the expressions for x and y, it affects only the size of the graph. The shape of the graph is determined by the parameter β. Thus, it is sufficient to set $a = 1$, and to consider the equations

$$x = t - \beta \sin t, \quad y = 1 - \beta \cos t. \tag{5}$$

Note that if $\beta = 1$, then $b = a$, and the trochoid becomes a cycloid.

(b) Consider values of β in the range $0 \leq \beta < 1$. If β is slightly less than 1, how do you think that the graph will be different from the cycloid (which corresponds to $\beta = 1$)? Now choose such a value of β, say $\beta = 0.99$, and draw the graph for two revolutions of the circle. Does the graph have the appearance that you expected?

To see the difference more clearly, you will have to look more closely at the neighborhood of a cusp. Thus you may want to restrict t to some relatively short interval about 2π.

(c) Now let β decrease further toward 0. Describe what you expect to happen to the maximum and minimum points on the graph as this occurs. Now plot the graphs for some values of β between 0 and 1, such as $\beta = 0.6$ and $\beta = 0.2$. Do the graphs agree with your expectations? Pay particular attention to the scale on the y-axis. You may want to specify the y range yourself in order to compare the graphs more easily. Finally, what is the situation when $\beta = 0$?

(d) Next, consider values of β that are larger than 1. How do you think that the graph will be different for such values? Choose $\beta = 1.01$, and plot the graph for two revolutions of the circle. Does the graph confirm your expectations? Again, you will probably need to restrict t to a neighborhood of a cusp. Then try some larger values of β, such as $\beta = 1.5, 2, 4$, and 6, to see what happens to the trochoid in these cases.

2. Rather than rolling a circle along a line, as in Problem 1, let us now suppose that a circle of radius b rolls without slipping on the inside of a larger circle of radius a. The path traced by a given point on the smaller circle is called a *hypocycloid*. To derive equations for this curve, suppose that the larger circle is centered at the origin, and that the given point on the smaller circle is initially located at the point (a, 0) where the larger circle intersects the positive x-axis. Then it is possible to show that a parametric representation for the hypocycloid is

$$x = (a - b) \cos \theta + b \cos\left(\frac{(a - b)\theta}{b}\right),$$

$$(6)$$

$$y = (a - b) \sin \theta - b \sin\left(\frac{(a - b)\theta}{b}\right) ,$$

where θ is the angle through which the smaller circle has been rotated.

(a) Try to derive Eqs. (6) yourself. Then let $\alpha = a/b$, and show that Eqs. (6) can be written in the form

$$x = b[(\alpha - 1) \cos \theta + \cos (\alpha - 1)\theta] , \qquad (7a)$$

$$y = b[(\alpha - 1) \sin \theta - \sin (\alpha - 1)\theta] . \qquad (7b)$$

Since b appears only as a multiplier in each of Eqs. (7), it affects only the size of the graph. Therefore let $b = 1$, and consider the equations

$$x = (\alpha - 1) \cos \theta + \cos(\alpha - 1)\theta , \qquad (8a)$$

$$y = (\alpha - 1) \sin \theta - \sin (\alpha - 1)\theta . \qquad (8b)$$

(b) Plot the graph of Eqs. (8) for $0 \le \theta \le 2\pi$, corresponding to one full circuit of the smaller circle around the larger circle, for $\alpha = 3$, for $\alpha = 4$, and for $\alpha = 5$. Describe what you think the graph will look like if α is a positive integer larger than 5 .

(c) Let $\alpha = 5/2$, and plot the graph for $0 \le \theta \le 2\pi$. Note that the curve does not close, as it did for the cases in part (b). This means that the point whose path is being traced did not return to its starting point after one full revolution. What can be done to close the curve? Try plotting it on the interval $0 \le \theta \le 4\pi$. Then let $\alpha = 7/2$ and plot the corresponding graph.

Based on these examples, can you make a statement about the graph for other values of α that are rational numbers, that is, fractions? Test your statement by plotting the graph for $\alpha = 7/3$.

(d) Describe how you would produce a starlike figure, first with eight points, and then with an arbitrary number n of points.

(e) Suppose that $\alpha = 17^{1/2}$. How does the graph in this case compare with the graph when $\alpha = 4$ that you found in part (b)? What will happen as you plot the graph over longer and longer time intervals? What is the situation if α is some other irrational number?

Note: **Be prudent in drawing graphs over long time intervals. This can be expensive in terms of time and memory.**

9

Rectangular Approximations and Riemann Sums

Objective

To demonstrate the use of rectangles to approximate the area under a curve, and to estimate the area by Riemann sums.

The definition of the definite integral is usually quickly forgotten once the Fundamental Theorem is discovered. This is understandable because the Riemann sums used in the definition are tedious to calculate by hand, and the Fundamental Theorem doesn't require taking a limit. However, with computers and Maple it isn't that difficult to calculate a Riemann sum, and in some cases it is possible to take the limit (see Lab 10). What you are asked to do in this Laboratory is to explore the use of Riemann sums. This will help in your understanding of the definition of the integral, and it will also show how the definition is useful in constructing approximations of the integral.

It is worth recalling the definition of a definite integral, which states that

$$\int_a^b f(x)\, dx = \lim_{\substack{n \to \infty \\ \|P\| \to 0}} \sum_{i=1}^{n} f(x_i^*)\, \Delta x_i \;,$$

where Δx_i is the length of the i-th subinterval; x_i^* is any point in the i-th subinterval; and $\| P \|$ is the length of the largest subinterval in the partition P of the interval [a , b] . The notation used here may differ slightly from that used in your text so you may want to review your textbook to help you understand this formula. In this Laboratory the subintervals have equal length and we investigate the consequences of making different choices for x_i^* . In particular, we investigate the case where x_i^* is taken to be the right endpoint of the i-th subinterval, then the left endpoint, and then the midpoint.

EXAMPLE

Consider the area between the graph of $y = 2 - \frac{1}{4} x^3$ and the x-axis for $0 \le x \le 2$. Approximate this area by using sets of inscribed and circumscribed rectangles. Use these rectangles to estimate the numerical value of the area.

To visualize the area in question, we enter the formula for f and then plot its graph over the interval:

```
> f := 2 − (x^3)/4;
> plot(f,  x = 0..2);
```

We are to approximate the area of the region that is bounded by the coordinate axes and by the graph shown in the plot. Rectangular approximations to the region are drawn by the Maple commands `leftbox` and `rightbox,` respectively, which reside in the Maple `student` package. The next step is to call up this package. We end the command with a colon so as to suppress the output.

```
> with(student):
```

The `leftbox` command has three arguments; the first is f , the second is the x interval, and the third is the number of subintervals, or rectangles, to be used in the approximation. The following command calls for 20 rectangles on the interval x = 0..2 , so each rectangle has a width of 0.1:

```
> leftbox(f, x = 0..2, 20);
```

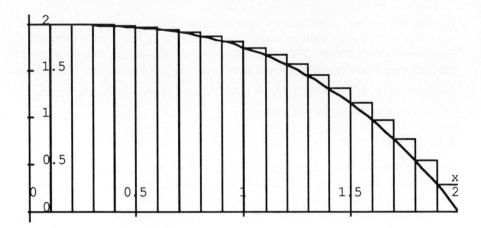

Each rectangle drawn by `leftbox` has height equal to the value of f at the left endpoint of the subinterval. Since f is a monotone decreasing function, this means that each rectangle extends above the graph. Thus the total area of the 20 rectangles is larger than the area under the curve (this is also apparent in the preceding figure). To find the total area of the 20 rectangles, and hence an upper bound for the area under the curve, we can use the command `leftsum`, whose syntax is the same as for `leftbox`:

> leftsum(f, x = 0..2, 20);

$$1/10 \left(\sum_{i = 0}^{19} (2 - 1/4000\ i^3) \right)$$

To obtain a numerical value for this sum, we use the `value` command:

> value(");

$$\frac{1239}{400}$$

Of course, to obtain a decimal number we use `evalf`:

> evalf(");

$$3.097500000$$

All of these commands can be combined in a single line, namely

> evalf(value(leftsum(f, x = 0..2, 20)));

We could obtain better estimates by using more rectangles and correspondingly smaller subintervals. However, first let us obtain a lower bound for the area by using inscribed rectangles. In the present problem these are drawn by `rightbox,` which plots rectangles whose heights are equal to the values of f at the right endpoints of the subintervals. The command `rightsum` then determines the total areas of the rectangles.

> rightbox(f, x = 0..2, 20);

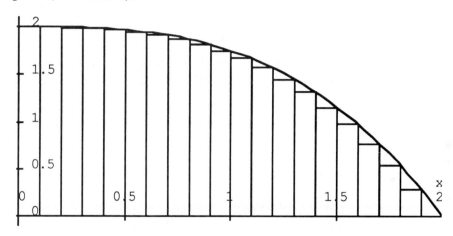

> evalf(value(rightsum(f, x = 0..2, 20)));

2.897500000

Maple also has a command, `middlesum,` that computes the areas of rectangles whose heights are the values of f at the midpoints of the subintervals. In this example, this will produce a value between those of `leftsum` and `rightsum.` (There is no corresponding `middlebox` command, so you will have to imagine the rectangles.)

> evalf(value(middlesum(f, x = 0..2, 20)));

3.001250000

The actual value of the area is 3. (You can check this if you already know how to evaluate the integral of f from x = 0 to x = 2 .) Thus the value given by `middlesum` is fairly accurate. Now let us see the effect of doubling the number of rectangles:

> evalf(value(leftsum(f, x = 0..2, 40)));

3.049375000

> evalf(value(rightsum(f, x = 0..2, 40)));

$$2.949375000$$

> evalf(value(middlesum(f, x = 0..2, 40)));

$$3.000312500$$

All of the results are more accurate, as expected. Note that the error in `leftsum` and `rightsum` has been reduced by about one-half, while the error in `middlesum` has been reduced by one-quarter.

Question: Will the same thing happen again if the number of rectangles is doubled again, to 80? Try it and find out.

To see the effect of steadily increasing the number of rectangles, one can write a loop that causes Maple to construct several plots in rapid sequence. For example:

```
> for n from 1 to 5 do
>    leftbox(f, x = 0..2, 10*n);
> od;
```

As noted before, the function f in this example is monotone decreasing, so `leftbox` and `leftsum` give an upper bound for the area, while `rightbox` and `rightsum` give a lower bound. If the function had been monotone increasing, then the situation would be reversed: `leftbox` and `leftsum` give a lower bound, while `rightbox` and `rightsum` give an upper bound.

As a final comment, there are two Labs related to this one that you may find interesting. One is Lab 11, which uses the ideas discussed here to find the area of a reasonably complicated region (namely the area of Pennsylvania), and in Lab 16 more sophisticated numerical approximations are explored.

Laboratory Problems

In the problems that follow n always refers to the number of approximating rectangles, or to the number of subintervals. Also, the three problems in this Lab are independent of one another, although Problem 1 serves as an introduction to the ideas in the Lab.

1. Let $f(x) = \sin x$, and consider the area under the graph of this function from $x = 0$ to $x = \pi/2$. The exact value of this area is 1.

    (a) Use `leftbox` and `rightbox` with 20 rectangles to plot rectangular approximations to this area. Which approximation is too large and which is too small? Why?

(b) Use `leftsum` and `rightsum` with 20 subintervals to calculate approximate values for the area.

(c) Repeat the calculations in part (b) with $n = 40$ and with $n = 80$. How do the errors in these approximations change as n increases?

(d) Evaluate `middlesum` for $n = 20$, 40, and 80. How do these approximations compare with those in parts (b) and (c)? How does the error change in this case as n increases?

(e) Find the smallest value of n for which the error in using `middlesum` is less than 0.000005. This question may be a little more difficult than it seems at first, since roundoff error may play a role. Determine the required value of n first using Maple's customary 10 digits. Then try 15 digits and find out if the answer is the same. Remember to reset Digits to 10 when you finish.

2. Let $f(x) = (\cos \sqrt{x})^2$, and consider the area under the graph of this function from $x = 0$ to $x = \pi^2/4$.

(a) Use `leftbox` and `rightbox` with $n = 20$ to plot rectangular approximations to this area. Which approximation is too large and which is too small? Why?

(b) Evaluate `leftsum` and `rightsum` for $n = 20$, 40, 80, and 160.

(c) Evaluate `middlesum` for $n = 20$, 40, 80, and 160. Do these values appear to be converging faster than those found in part (b)?

(d) By using larger values of n in `middlesum`, try to determine a value that you are confident gives the area correct to five decimal places. One way to do this, without knowing the exact value of the area, is to increase n and note how many decimal places remain unchanged.

(e) Try to determine whether the values of `middlesum` are larger or smaller than the actual area.

3. Let $f(x) = \exp(x) \sin(x)$, and consider the area under the graph of this function from $x = 0$ to $x = \pi$.

(a) Plot the graph of this function. Note that it is increasing for part of the interval and decreasing for the rest.

(b) Use `leftsum` and/or `rightsum` to calculate an upper bound for the area under this curve, using a total of 40 subintervals. In a similar way, calculate a lower bound for the area. Repeat with $n = 80$ and $n = 160$.

(c) Use `middlesum` with suitable values of n to determine an approximate value for the area that you are confident is correct to three decimal places (five digits).

10

The Fundamental Theorem of Calculus

Objective

To emphasize the definition of the definite integral as a Riemann sum and to use Maple to discover the Fundamental Theorem of Calculus.

The purpose of this Laboratory is to help you understand how the definition of the definite integral as a limit of Riemann sums leads to the Fundamental Theorem of Calculus. The best place to begin is by recalling the definition of a definite integral, which is that

$$\int_a^b f(x)\,dx \;=\; \lim_{\substack{n\to\infty \\ \|P\|\to 0}} \sum_{i=1}^n f(x_i^{*})\,\Delta x_i \;,$$

where Δx_i = length of the i-th subinterval; x_i^{*} is some point from the i-th subinterval; and $\|P\|$ = length of the largest subinterval in the partition P of the interval $[a, b]$. The notation used here may differ slightly from what is used in your textbook, so you may want to review your text to help you understand the formula.

We will see how this definition can be used to discover one of the most important theorems in calculus.

Laboratory Problems

The problems in this Lab are designed to lead to certain conclusions when the Lab is completed. Therefore, even though some of the individual problems are independent, the Laboratory as a whole relies on all the problems.

1. Consider the function $f(x) = x^3$ over $[a, b]$ and a partition P of $[a, b]$ into n equal subintervals.

    (a) What is the length of the i-th subinterval? Define this length as dxi in Maple.

    (b) What is the rightmost point of the i-th subinterval? Define this point as xi in Maple.

    (c) Use Maple to find an expression for the Riemann sum corresponding to this partition and taking x_i^* to be the rightmost point, xi, in each subinterval.

    (d) Find the limit of this expression as $n \to \infty$. You should get $\frac{1}{4}b^4 - \frac{1}{4}a^4$.

    (e) Write down an antiderivative $G(x)$ for x^3, and comment on how it compares with the value of the limit in (d).

2. Now consider $f(x) = x^3 + 5x$ over $[a, b]$ and a partition P of $[a, b]$ into n equal subintervals.

    (a) Use Maple to find an expression for the Riemann sum corresponding to this partition. As before, take x_i^* to be the rightmost point, xi, in each subinterval.

    (b) Find the limit of this expression as $n \to \infty$.

    (c) Write down an antiderivative $G(x)$ for $x^3 + 5x$, and comment on how it compares with the value of the limit in (b).

3. As one final example, consider the function $f(x) = \cos(x)$ over $[a, b]$ and a partition P of $[a, b]$ into n equal subintervals.

    (a) Use Maple to find an expression for the Riemann sum corresponding to this partition. As before, take x_i^* to be the rightmost point, xi, in each subinterval.

    (b) Find the limit of this expression as $n \to \infty$.

    (c) Write down an antiderivative $G(x)$ for $\cos(x)$, and comment on how it compares with the value of the limit in (b).

4. In each of the preceding three problems, how does the limit of the Riemann sum appear to be related to the antiderivative $G(x)$? You may answer this question by filling in the right hand side of the following equation:

$$\int_a^b f(x)\, dx \;=\; \lim_{n \to \infty} \sum_{i=1}^{n} f(x_i)\, dx_i \;=\; \underline{\hspace{4in}}$$

(An expression involving G, a, and b.)

5. For the function $f(x) = x^3$ studied in Problem 1, show that the limit of the Riemann sum remains unchanged if we pick

$$x_i = \text{the leftmost point in the i-th subinterval,}$$

or if we pick

$$x_i = \text{the midpoint of the i-th subinterval.}$$

11

Integral Approximations to an Area

Objective
To construct and to evaluate integral approximations for the area of a region that is specified by a set of data points.

The most basic interpretation of the integral

$$\int_a^b f(x)\, dx$$

is that its value is the area between $y = f(x)$ and the x-axis for $a \le x \le b$. This Laboratory takes advantage of this interpretation by using simple approximations of integrals to calculate the area of a region. The problem is to calculate approximately the area of a state (like Pennsylvania) given some limited information about its boundaries. You will construct several approximations to the area, corresponding to different integral approximations. In the course of the calculations, you will see how to manipulate data points using Maple. Finally you will see effects of improved integral approximations on the estimated value of the state's area. The use of Riemann sums in approximating, or evaluating, an integral also arises in labs 9, 10, and 16.

EXAMPLE

In this Laboratory you will use the summation (`sum`) command, and the integral approximations `leftsum` and `simpson,` in the Maple `student` package, are mentioned but not used. Also, because of the nature of this Laboratory you will have to be able to manipulate data points. To illustrate how this is done, suppose we need to work with the following three points in the x,y-plane: $(1 , 5), (2 , 3), (-1 , 4)$. To enter the x-coordinates into Maple, we use an array by entering the command

> x := array(0..2, [1, 2, −1]);

The first argument of the command tells Maple that x is an array (or vector in this case) with three elements, and the subscripts for these elements are 0 , 1 , and 2 . The second argument is the actual data (note the square brackets). We can then refer to individual elements of this array since $x[0] = 1$, $x[1] = 2$, and $x[2] = -1$.

Laboratory Problem

We would like to find an approximation to the total area ATOT of Pennsylvania (PA). A sketch of the state is shown at the end of the Lab. To begin, we need a strategy for formulating the problem mathematically and then solving it. A good strategy is one that is both fairly simple *and* allows us to take advantage of Maple's capabilities.

PART 1

(a) One approach is to first compute the area of PA that is west of the dashed line shown on the sketch. Then the area that is east of the dashed line is computed, after first rotating that region by 90 degrees in a counterclockwise direction. Explain why you think this strategy is a good one.

(b) Using the distances shown on the sketch and elementary geometry, compute the entire area west of the dashed line. Call this area AW . Note that except for the "Erie Chimney," the northern and southern boundaries of PA are approximated here as parallel lines. Also, the dashed line is perpendicular to the southern and northern boundary lines. The western boundary is approximated as a straight line.

Call the area east of the dashed line AE . To compute AE , we need an approximation for the eastern PA boundary, which for the most part is a river that George Washington once had to cross. Imagine AE rotated as described in (a). Assume an x,y-coordinate system with x-axis along the dashed line and origin at the point marked P . The information that is given about the eastern boundary consists of coordinates of 25 points, numbered 0 through 24, shown on the sketch.

(c) The x-coordinates should be in a Maple array named xpa , and the y-coordinates in a Maple array ypa . Enter this data with the commands

> xpa := array(0..24 , [0, 4, 8, 10, 26, 35, 38, 42, 45, 62, 68, 77, 84,

>   98, 98, 109, 109, 117, 126, 133, 141, 145, 152, 152, 158]);

> ypa := array(0..24 , [17, 22, 22, 30, 33, 37, 50, 53, 48, 37, 31, 36,

>   29, 29, 34, 38, 41, 47, 55, 44, 34, 34, 19, 5, 0]);

As an alternative, your instructor may provide instructions so you can read in these coordinates more easily (e.g., from a file located on the computer system). Note that you may refer to elements of the xpa and ypa arrays using brackets. For example, if you enter the command xpa[6]; then Maple responds with 38.

Plot the boundary points using the command:

> plot([[xpa[i], ypa[i]] $ i=0..24], style=POINT, title=`Boundary Points`);   (1)

Now plot the boundary points with straight lines connecting the points using the same command as in (1) but replacing POINT with LINE. You should also modify the title.

PART 2

(a) In order to find the area AE , we consider approximations to this geometric area. Let us start with the simplest approximation: a single rectangle. Choose what you think is a good rectangle to approximate AE , and briefly explain the reason for your choice. Use the area of this rectangle to obtain your first estimate of ATOT .

(b) We could try to select more rectangles to approximate AE , but the choices become increasingly arbitrary. Let us be more systematic. Suppose we use three rectangles with heights given by the y-coordinates at points 0 , 8 , and 16 . The mathematical formula for the area AE3i for the i-th rectangle of this set, i = 1 , 2 , 3 , is

$$AE3i = ypa\big(8(i-1)\big)\,\big(\,xpa(8i) - xpa\big(8(i-1)\big)\,\big)\ . \tag{2}$$

Express this formula as a Maple command and then evaluate this approximation using the command

> AE3 := sum(AE3i, i = 1..3);   (3)

In a corresponding way use six rectangles and compute an approximation AE6 by slightly modifying (2) and (3).

(c) Use the same process as in (b) and compute two more approximations AE12 and AE24 . Compute four estimates for ATOT using your values for AE3 , ..., AE24 . Why are your estimates analogous to those obtained if the Maple command `leftsum` is used? Can we use `leftsum` in this problem?

PART 3

(a) There is another natural type of approximation that can be applied to determine AE . If you select a few pairs of boundary points, and draw straight lines between them, in this way AE can be regarded as approximated by a sum of trapezoids. Do so by constructing four "good" trapezoids (of your choice), and find the area. Recall that the area of a single trapezoid is the average of its parallel sides times the perpendicular distance between these sides.

(b) Picking more trapezoids and entering coordinates as in part (a) becomes tedious. Imagine three trapezoids using equal segments along the x-axis, and let the area of the i-th trapezoid be denoted by AET3i , for i = 1 , 2 , 3 . Find the formula for AET3i by slightly modifying Eq. (2). Then use the Maple sum command to compute AET3 , the three-trapezoid estimate for AE .

(c) Modify your formula in part (b) to compute new estimates AET6 , AET12, and AET24 using 6 , 12 , and 24 trapezoids. Find the corresponding estimates for ATOT , including one for AET3 .

PART 4

(a) Other integration methods, such as a version of Simpson's method in the Maple student package, could be used for the approximation of AE . For the specific data provided, give at least one reason why Simpson's method may be difficult to apply.

(b) According to an encyclopedia, the area ATOT for PA is 45,333 square miles. Discuss the accuracy of your estimates for ATOT compared with this value.

(c) What aspects of our mathematical formulation for the area of PA do you think might account for discrepancies in answers in part (b)? What do you think has the largest effect? Be specific.

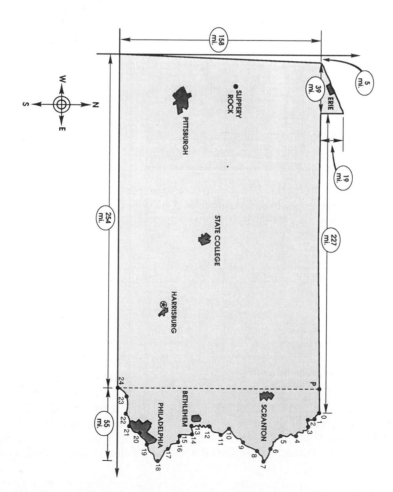

12

Applications of Solids of Revolution: Gas and Burgers

Objective

To use Maple to calculate the volume of a solid of revolution and then use this to design a gas gauge and find the volume of a hamburger.

The problems in this Laboratory involve calculating the volume of a solid of revolution. You are asked to use Maple to find the volume and then use the result to answer a question or two about the problem.

EXAMPLE

Consider the bowl shaped surface you get by rotating the curve

$$y = x^2 - 4, \text{ for } 0 \le x \le 2,$$

around the y-axis. You can get an idea of what the cross section of the bowl looks like by using the command

> plot(x^2 – 4, x = –2..2, title = `Plot of Cross Section of Bowl`);

Suppose the bowl contains a fluid (say, milk). The depth d of milk will be measured at the center of the bowl, so $0 \le d \le 4$. To determine the volume of milk note the limits on y are $-4 \le y \le -4 + d$. The volume is going to be determined here using the disk method (also called the method of slices), and the radius of the

108

disk is $r = x$, that is, $r = \sqrt{y + 4}$. The volume dV of the disk is πr^2 dy, and so, the Maple command that determines the volume is

> vol := int(Pi*(y + 4) , y = −4..−4 + d);

Now suppose we want the relative volume. This means we must calculate the volume when the bowl is full, and this is obtained using the command

> full := subs(d = 4, vol);

From this the relative volume is

> v := vol/full;

Laboratory Problems

The two problems in this Lab involve applications of solids of revolution, and these problems are independent of one another.

1. Have you ever wondered how the gas gauge in a car works? It determines the relative volume of gasoline in the tank by making a measurement of a variable like depth or pressure. In this problem you are asked to determine the relative volume of fluid in a tank as a function of depth. You are to also find an approximation of the volume and then discuss the consequences of using this approximation to determine 'E' and 'F' on a gas gauge.

    For the gas gauge, the variable of interest is

    $$v := \frac{\text{volume of fluid in tank}}{\text{volume of tank when full}} .$$

    So, v is the relative volume in the tank when the fluid is at a depth d (see figure).

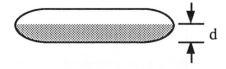

    (a) Suppose the tank is a volume of revolution obtained by rotating the curve shown here around the y-axis. Determine v as a function of d.

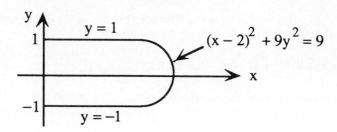

(b) Plot v as a function of d .

(c) It is generally much easier to construct measuring devices, like a gas gauge, that are based on linear functions. The problem here is that the dependence of v on d is nonlinear. However, as is apparent from your plot in (b), this function is approximately linear over a reasonably large range of d values. Use a suitable tangent line to find a linear approximation to v , and denote this approximating function as w . On the same graph, plot v and w as functions of d .

(d) Presumably you would not like to run out of gas when the gas gauge says the tank is not empty. At the same time you don't want to have to stop to get gas way before it is really necessary. Comment on these situations if one were to use the linear approximation in (c) for the gas gauge. In this discussion assume the gauge reads 'F' if $w \geq 1$ and it reads 'E' when $w \leq 0$. Do you feel this is the best linear approximation that can be made for this application, or is there a better one? If there is, describe what it is.

2. This problem attempts to answer the question of which restaurant has the biggest hamburger (or veggie burger). To do this we will assume the hamburger can be described as a solid of revolution. In particular, we will assume that the hamburger is formed by rotating the curve shown here around the y-axis (the curve includes the upper portion where $y = f(x)$ as well as the vertical piece where $x = a$).

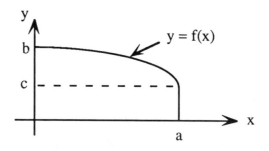

The function describing the top of the bun is $y = f(x)$, for $0 \le x \le a$, where

$$f(x) = b - Ax^n \ .$$ (1)

Because it is required that $y = c$ when $x = a$, then in the formula it must be that

$$A = \frac{b - c}{a^n} \ .$$ (2)

So given a hamburger, we must measure the radius a, the thickness b at the center, and the height c. We must also determine the value of n, which measures the "steepness" of the bun.

To answer the questions that follow, it is recommended that you measure the values of a, b, and c for a hamburger at your favorite eatery. In lieu of doing this, you can use the values $a = 2.4$, $b = 2$, and $c = 1.3$ (these were obtained from a hamburger from a well-known national chain, and the units here are inches).

(a) Using your values of a, b, and c plot f for various values of n to determine what n should be. To do this it may be necessary to resize the plot window after each plot so the bun is not distorted. Plot your choice for the hamburger, and in your write-up state what requirement(s) you used to determine n.

(b) Calculate the volume of the hamburger.

(c) How significant is the fact that the bun is curved? Investigate this by calculating the volume for a flat bun ($f = b$) and comparing it with your value from (b).

13

Applications of Solids of Revolution: Equal Volumes

Objective

To use Maple to assist in dividing solids of revolution into pieces of equal volumes.

Did you ever try to divide a freshly baked pie into seven equal pieces? You know the "theoretical" solution, that you need to make seven cuts so that the vertex angle of each wedge is between 51 and 52 degrees! You might consider taking an easy way out, by dividing the pie into eighths and eating two pieces yourself. However, this solution violates what is known as the Principle of Fair Parenthood: Every family member gets an equal share of the pie.

Consider the application of the Principle of Fair Parenthood to the problem of slicing fruits to yield pieces of equal volumes. We limit ourselves to oranges and watermelons, and to parallel slices.   The problem can then be reduced to one involving an application of solids of revolution.   Similar problems were also considered in the previous Lab.

EXAMPLE

(a) We model an orange as a sphere with radius R . Also, suppose a cap of thickness w is sliced from the orange (the figure following depicts a cross section of this sliced orange).

To determine the volume of the slice as a function of R and w , imagine the whole orange (and the slice) as a solid of revolution. The whole orange is formed by rotating a semi-circle about the x-axis:

> y := sqrt(R^2 − x^2);

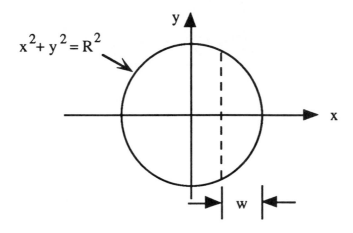

The volume of the whole orange is therefore

> V := Pi*int(y^2, x = −R..R);

This result is familiar, of course. We can use the same idea to calculate the volume of the cap. This time the volume extends only from the location of the cut (at x = R − w) to the end of the orange (x = R):

> Vw := Pi*int(y^2, x = R − w..R);

Notice the familiar answers when w = R and w = 2R .

(b) Where should we cut the orange to produce two pieces of equal volume?

(c) Suppose the orange has radius R = 1 . Where should we cut it to produce three pieces of equal volume? By symmetry we should cut the orange at some point x = p , where 0 < p < 1 , and **also** at x = − p . The value of p can be determined by setting the volume of one of the two caps equal to 1/3 the volume of the whole orange. Letting Vp denote the volume of the cap, we then enter the equation we want to solve:

> Vp := Pi*int(y^2, x = p..R);
> eq := subs(R = 1, Vp) = subs(R = 1, V)/3;

This equation can be simplified and the Pi removed:

> eqs := simplify(eq/Pi);

This is a cubic equation for p. We can solve for its roots:

> roots := fsolve(eqs, p);

$$roots := -1.833986597, .2260737138, 1.607912883$$

The only root of interest is the one between 0 and 1:

> P:=roots[2];

$$P := .2260737138$$

Therefore, two cuts, each about 22.6% away from the center section of the orange, will divide it into thirds.

Laboratory Problems

In this Lab you are to divide oranges (in Problem 1) and watermelons (in Problem 2). The problems are independent.

1. (a) For the orange with radius $R = 1$, where should we cut to produce

   (i) four pieces of equal volume?

   (ii) five pieces of equal volume?

   (b) Suppose we want to divide oranges of **different** sizes into five pieces of equal volume.

   (i) First, set up equations as in part (a) but for an arbitrary radius R. What kind of trouble arises if we try to use `fsolve` directly? If we try to use `solve`? This means we need a better problem formulation.

   (ii) To help in finding a better formulation, consider a particular case: for example, take $R = 10$. Where should the cuts be made? In view of your answers for $R = 1$ and $R = 10$, where should the cuts be made for $R = 2$? For *any* value of R? Justify your conclusion.

2. (a) Let's model a watermelon as an ellipsoid. That is, the melon is formed by rotating an ellipse about its major axis (which we take as the x-axis). Write the equation of the ellipse as

$$y = \text{sqrt}(1 - e\char`\^2)*\text{sqrt}(1 - x\char`\^2) \,,$$

where e is a parameter called the eccentricity (it is left unspecified in this problem). We want to cut the melon in planes perpendicular to the major axis to produce three pieces of equal volume.

   (i)   Where should the cuts be made?

   (ii) Compare the locations of the cuts with the corresponding points for the orange (see the example, part (c)). Try to explain your result.

(b) Another model for a watermelon is the volume of revolution generated by rotating the curve

$$y = 0.475*\text{sqrt}(1 - x \char`\^ 4)$$

about the x-axis.

   (i)   Plot a cross section of this model, and compare it with a plot of the model used in part (a) of the example with $e = 0.88$.

   (ii) Where should we cut this melon to produce three pieces of equal volume? Compare the locations of the cuts with those in part (c) of the example. Are your results sensible? Justify your conclusion.

14

Polygonal Approximation of Arc Length

Objective

To investigate the derivation of the arc length formula using a polygonal approximation, and to examine how this approximation can be used to compute arc length.

In this Laboratory you are to investigate the polygonal approximation used to derive the formula for arc length. The usual procedure in this derivation begins by calculating the lengths of the straight lines forming the polygon. By considering better and better polygonal approximations of the curve one ends up with a formula for arc length that involves the derivatives of the parametrization. The derivation of this formula is not particularly easy to follow, and the purpose of this Lab is to use Maple to investigate some of the steps involved.

To help explain the procedure we will briefly describe the principal ideas underlying the polygonal approximation. You are encouraged to also look over the discussion in your textbook.

Suppose we are given a curve with the parametric representation

$$\left. \begin{array}{l} x = f(t) \\[2mm] y = g(t) \end{array} \right\} \quad \text{for } 0 \le t \le \ell .$$

To obtain a polygonal approximation of the curve we will use a uniform partition of the t interval. So, the i-th point in the partition is $t_i = ih$ for $i = 0, 1, 2, ..., n$, where $h = \ell/n$ (see the following figure). Each point t_i of this partition determines a point on

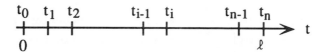

the curve with coordinates (x_i, y_i). As an approximation of the arc length between the points (x_{i-1}, y_{i-1}) and (x_i, y_i), the length of the straight line connecting these points is used (see following figure).

Adding these lengths together, we are led to the following approximation for the arc length:

$$\Delta L = \sum_{i=1}^{n} \sqrt{(x_i - x_{i-1})^2 + (y_i - y_{i-1})^2} \ . \tag{1}$$

If the curve is smooth enough (i.e., it is rectifiable) we will have

$$\lim_{n \to \infty} \Delta L = L \ .$$

Here L is the arc length of the curve, and in most calculus textbooks it states that

$$L = \int_0^{\ell} \sqrt{(f')^2 + (g')^2} \ dt \ . \tag{2}$$

In this Laboratory we are going to investigate some of the connections between (1) and (2).

EXAMPLE

To explain how to use Maple to determine ΔL, consider the curve

$$\left.\begin{array}{l} x = \cos(t) \\[2mm] y = \sin(t) \end{array}\right\} \quad \text{for } 0 \leq t \leq \pi. \tag{3}$$

Suppose we use a partition with $n = 4$ (so $h = \pi/4$). In this case $t_0 = 0$, $t_1 = \pi/4$, $t_2 = \pi/2$, $t_3 = 3\pi/4$, and $t_4 = \pi$. To plot the polygonal approximation we first calculate the points on the curve determined by the partition. This is done using the following Maple commands (recall $t_j = hj$):

```
> h := Pi/4;
> points5 := [ [ cos(h*j), sin(h*j) ] ] $ j = 0..4 ];
```

We now draw these points and connect them by straight lines. On the same plot we include the plot of the original curve:

```
> plot( { points5, [ cos(t), sin(t), t = 0..Pi ] }, style = LINE, title = `EXAMPLE` );
```

The word LINE must be in capital letters in this command (the title for the plot is in capital letters simply so it will be easier to read when printed).

The next step is to calculate ΔL . To do this we need to first calculate the length of the individual line segments, that is, the square root in (1). This is accomplished with the following sequence of commands:

```
> xd := i -> cos(i*h) – cos((i – 1)*h);
> yd := i -> sin(i*h) – sin((i – 1)*h);
> dist := i -> abs( sqrt( xd(i)^2 + yd(i)^2 ) );
```

$$dist := i \rightarrow abs(sqrt(xd(i)^2 + yd(i)^2))$$

Therefore dist(i) is the length of the line segment from the (i–1)-st point to the i-th point. The abs($\cdot$) is included in the formula for the distance function to guarantee that Maple uses the nonnegative square root. The approximation for the arc length in (1) is now determined by carrying out the summation. This is done using the `sum` command as follows:

```
> dL := sum( dist(i) , i = 1..4 );
```

$$dL := 3\,((1/2\,\,2^{1/2} - 1)^2 + 1/2)^{1/2} + (1/2 + (1 - 1/2\,\,2^{1/2})^2\,)^{1/2}$$

To determine the actual value for the approximation we have Maple evaluate the terms in the sum using the command

```
> evalf( dL );
```

3.061467459

So, the value for dL differs from the exact value of π by less than 3%. You might want to compare the exact result for larger values of n (say, $n = 8$, 16, and 32).

Laboratory Problem

1. (a) Explain why $\Delta L \leq L$ irrespective of the partition that is used.

Consider the specific example of a hypocycloid of three cusps (or, deltoid), where

$$\left. \begin{array}{l} x = 2\cos(t) + \cos(2t) \\[2mm] y = 2\sin(t) - \sin(2t) \end{array} \right\} \quad \text{for } 0 \leq t \leq 2\pi . \qquad (4)$$

(b) Use (2) to calculate the arc length of this curve.

(c) Plot the curve. On this plot also draw the straight lines used to determine ΔL in the case of when $n = 3$.

(d) Plot the curve, and the straight lines used to determine ΔL, in the case of when $n = 6$.

(e) Plot the curve, and the straight lines used to determine ΔL, in the case of when $n = 12$.

(f) Plot the curve, and the straight lines used to determine ΔL, in the case of when $n = 24$.

(g) Calculate ΔL for $n = 3, 6, 12, 24$. Based on these values does it appear that $\Delta L \rightarrow L$ as n increases? Is your conclusion consistent with the plots you obtained in (c) – (f)?

(h) For this curve explain why one should take n to be a multiple of 3. Illustrate your answer by comparing the value of ΔL for $n = 6, 7, 8, 9, 10$. In particular, calculate the error, Error $:= L - \Delta L$, for these values of n.

EXTENSION OF LABORATORY PROBLEM

The polygonal approximation furnishes us with a way to evaluate the arc length integral. This is worth considering since most integrals for arc length are difficult and have to be evaluated numerically. There are, however, other ways to do this and possibilities include the rectangular approximations discussed in Lab 9. What we are interested in here is how well the polygonal approximation does in comparison with these other methods. So, for the following curves, compare the value of the arc length integral obtained using rightsums, middlesums, and the polygonal approximation, in the case of when $n = 6$, 12, 18, and 24.

(i)   The curve $y = \exp(x)$ for $-1 \leq x \leq 1$.

(This can be written in parametric form as $x = t$, $y = \exp(t)$ for $-1 \leq t \leq 1$.)

(ii)   The ellipse

$$\left. \begin{array}{l} x = 2\cos(t) \\ \\ y = 3\sin(t) \end{array} \right\} \quad \text{for } 0 \leq t \leq 2\pi .$$

(iii)   The deltoid given in (4).

15

Application of Integrals for Arc Length: Travel Time

Objective
To use Maple for calculating integrals along curves and to illustrate how the time to ski down a hill is affected by its shape.

How much time does it take for a skier to ski down a slope? A simple question, but a precise answer depends on many things. For instance, does the skier push hard all the way down, or just glide down under gravity after a gentle initial push? Is the skier experienced, with a good form all the way down? Along the way, does the skier "wipe out"? Does the skier go straight down the slope, or cut across it now and then? Is the wind blowing, and if so which direction? How many other factors can you name that might have an effect? We consider the influence of just **one** factor, the shape of the slope. We assume the skier is "skilled but lazy," so that the skier glides straight down the hill under gravity with no pushes and does not fall.

Suppose the top of the hill is at the point $(0, 1)$ in the x,y-plane, and the bottom is at $(\pi/2, 0)$. Different shapes of the hill correspond to different curves connecting these two points. We let the shape of the hill be given by the parametric equations $x = f(u)$ and $y = g(u)$ for $0 \le u \le \pi$. These functions are entered into Maple as follows:

```
> x := u -> f(u);
> y := u -> g(u);
```

The bottom of the hill at $x = \pi/2$ and $y = 0$ (which both occur when $u = \pi$) is chosen to simplify some of the parametric equations to follow.

This Laboratory contains an application of arc length integrals. Some of the ideas underlying the derivation of the arc length formula are discussed in lab 14.

EXAMPLE

(a) Explain why the length of the ski slope is given by

```
> L := int( sqrt( diff(f(u), u)^2 + diff(g(u), u)^2 ) , u = 0..Pi );
```

(b) Now we find the elapsed time T for the skier. The principle of conservation of energy means that the sum of the kinetic and potential energies of the skier is a constant. This is because we are neglecting energy losses (like friction between the skis and the slope) and energy gains (like pushes). Thus

$$\frac{1}{2}mv^2 + mGy = k \ , \tag{1}$$

where v and m are the skier's speed and mass, G is acceleration due to gravity, and k is a constant. (We use G for gravitational acceleration since g is used for a parametric equation.) If the skier starts off with zero velocity (a very lazy skier indeed), show that $k = mG$. Substitute this value for k into Eq. (1) and solve for v. Then replace v by ds/dt, where s is arc length along the hill, and show that

$$dt = \frac{d\,s}{\sqrt{2G(1-y)}} \ .$$

Finally, express the arc length differential ds in terms of f' and g', and integrate with respect to u, to give the following expression for T:

```
> T:=(2*G)^(−1/2)*int(sqrt((diff(f(u), u)^2 + diff(g(u), u)^2)/(1 − g(u))), u = 0..Pi);
```

(c) The simplest shape for the slope is a straight line. Remember that the hill must connect the points $(0 , 1)$ and $(\pi/2 , 0)$. For this hill, the equation is $y = 1 - 2x/\pi$. What are the parametric equations? For $u = 0$, x has to be 0; and for $u = \pi$, x has to be $\pi/2$. Therefore one choice for $f(u)$ is

```
> f(u) := u/2;
```

For this choice of f, it follows that

```
> g(u) := 1 − u/Pi;
```

As a check notice that $g(0) = 1$ and $g(\pi) = 0$, as it should be. Now calculate the length L and elapsed time T for this hill:

> L; T;

You should find that $L = \text{sqrt}(1 + (Pi/2)^2)$ and that $T = \text{sqrt}(2/G)$ times L . Does the first make sense considering the geometry? Can you get the second answer integrating by hand? The calculations for L and T in the example are easy because the shape of the hill is so simple, so you do not need Maple. However, the assistance of Maple is **essential** for the calculation of L and T for other hill shapes.

Laboratory Problem

1. For each of the following hill shapes, calculate the length L and elapsed time T . If the hill is not given in parametric form, first find appropriate parametric equations. [Remember that u should vary between 0 and π, and that f should be chosen to satisfy $f(0) = 0$ and $f(\pi) = \pi/2$ while g should satisfy $g(0) = 1$ and $g(\pi) = 0$.]

   (a) $f(u) = u/2$, $g(u) = 1 - \sin(u/2)$,

   (b) $f(u) = (u - \sin(u))/2$, $g(u) = (1 + \cos(u))/2$,

   (c) $x = (\pi/2)(1 - y)^2$,

   (d) $y = 1 - (2x/\pi)^2$,

   (e) $f(u) = (\pi/2)(1 - \cos(u/2))$, $g(u) = 1 - \sin(u/2)$.

   What conclusions can you draw from your calculations? List the set of curves in order of increasing length L . List them in order of increasing elapsed time T . How do the two lists differ? Plot some or all of the curves; do the pictures help to explain the orders of the lists?

16

Numerical Integration

Objective
To use Maple to develop an understanding of how to use an approximation of a
definite integral.

In many real life problems it is almost inevitable that definite integrals arise
that cannot be evaluated using the Fundamental Theorem. In such cases one is
faced with having to construct an approximation to the value of the integral, and this
Laboratory investigates how to do this.

The simplest approximation is to use the Riemann sums that arise in the
definition of the definite integral. To explain this and introduce the notation for the
Laboratory, suppose we want to evaluate the definite integral

$$\int_a^b f(x)\, dx \ .$$

To construct the approximation we divide the interval $[a, b]$ into n equal
subintervals, with endpoints $x_i = a + ih$ for $i = 0, 1, 2, ..., n$. The approximation
using Riemann sums is then

$$\int_a^b f(x)\,dx \approx h \sum_{i=1}^{n} f(x_i^*) \ , \tag{1}$$

where x_i^* is a point from the i-th subinterval, and $h = (b - a)/n$. The error, which we will denote as E_R , in this approximation satisfies

$$|E_R| \le \frac{(b-a)^2}{2n} M_1 \ , \quad \text{where } M_1 = \max_{a \le x \le b} |f'(x)| \ . \tag{2}$$

The two other approximations we will consider in this Laboratory are the following:

Trapezoidal:

$$\int_a^b f(x)\,dx \approx \frac{h}{2}[f(a) + f(b)] + h \sum_{i=1}^{n-1} f(x_i) \ ,$$

and the error satisfies

$$|E_T| \le \frac{(b-a)^3}{12n^2} M_2 \ , \quad \text{where } M_2 = \max_{a \le x \le b} |f''(x)| \ . \tag{3}$$

Simpson:

$$\int_a^b f(x)\,dx \approx \frac{h}{3}[\,f(x_0) + 4f(x_1) + 2f(x_2) + \ldots + f(x_n)] \ ,$$

and the error satisfies

$$|E_S| \le \frac{(b-a)^5}{180n^4} M_4 \ , \quad \text{where } M_4 = \max_{a \le x \le b} |f''''(x)| \ . \tag{4}$$

You should refer to your textbook for an explanation of these formulas and how they are derived. Also, you may find Labs 9 and 11 of interest because they deal with using Riemann sums to approximate integrals.

EXAMPLE

To demonstrate how to use Maple to calculate the preceding approximations, we will consider the definite integral

$$\int_0^1 \exp(x^3)\,dx \ .$$

First we have Maple evaluate it so we can compare the result with the approximations to come later:

```
> f := exp(x^3);
> int(f, x = 0..1);
> evalf(");
```

> > > > > > > > *1.341904418*

The approximations are in the `student` package. So, let's see how well the trapezoidal approximation does if four subintervals are used:

```
> with(student):
> trapezoid(f, x = 0..1, 4);
```

$$1/8 \ + \ \frac{1}{4}\left(\sum_{i=1}^{3} exp(\ 1/64\ i^{3}\ )\right) + 1/8\ exp(1)$$

```
> evalf(value("));
```

> > > > > > > > *1.383213747*

Note that in the last step we used a composite command to evaluate the approximation. To compare let's try Simpson's approximation:

```
> simpson(f, x = 0..1, 4):
> evalf(value("));
```

> > > > > > > > *1.345570102*

So, it appears that Simpson's approximation wins this contest (i.e., it is more accurate, when $n = 4$, than the trapezoidal method). This is not unexpected because of the bounds on the errors that were given. The n^4 in the denominator in (4) means that we can generally guarantee that the error for Simpson is smaller, for large n, than the error for the others. This observation, and the inequality in (2), also indicate why Riemann sums do not usually furnish good approximations. To see if this is the case for our example, let's take x_i^* to be the right endpoint of the i-th subinterval. The Maple commands to calculate (1) are:

```
> evalf(value(rightsum(f, x = 0..1, 4)));
```

> > > > > > > > *1.597998975*

As expected we do not have as accurate a result as with the other two methods. This is one of the reasons that Riemann sums are rarely used to evaluate an integral.

Laboratory Problems

The two problems of this Laboratory are independent of one another. In Problem
2(b) a reference is made to Problems 1(b), (c). However, it is not necessary to do
the problems in Problem 1 to understand, or be able to do, what is asked for in
Problem 2.

1. We will apply the preceding approximations to evaluate

$$\int_0^{10} \exp(\sin(x)) \, dx \; .$$

   In this case, $f(x) = \exp(\sin(x))$.

   (a) Have Maple evaluate the integral (call this value A).

   (b) Plot f'' for $0 \le x \le 10$. You do not need to hand this plot in but use
   it, along with `fsolve` and f''', to find M_2 given in (3).

   (c) Using ideas similar to those described in (b), find M_4 given in (4).

   (d) Use (3) to determine how large n should be to guarantee that the error
   in the trapezoidal approximation is no more than 10^{-4}. Check this by
   calculating the trapezoidal approximation for $n = 4, 8, 16, 32, 64$
   (label these approximate values T4 , T8 , T16 , T32 , and T64). The
   inequality in (3) suggests that the error should be cut by a factor of 4
   if we double n. Calculate $(A - T4)/(A - T8)$, $(A - T8)/(A - T16)$,
   $(A - T16)/(A - T32)$, and $(A - T32)/(A - T64)$. After this comment
   on whether or not the decrease occurs.

   (e) Use (4) to determine how large n should be to guarantee that the error
   in the Simpson approximation is no more than 10^{-4}. Check this by
   calculating Simpson's approximation for $n = 4, 8, 16, 32$ (label
   these values S4 , S8 , S16 , S32). The inequality in (4) suggests that
   the error should be cut by a factor of 16 if we double n. Calculate the
   ratios $(A - S4)/(A - S8)$, $(A - S8)/(A - S16)$, and $(A - S16)/(A - S32)$. Comment on whether or not the decrease occurs.

2. In this problem you are to find an accurate approximation to the integral
   that determines the period of a pendulum. In preparation for this, let's
   recall a few facts about pendulums (see the following figure). As shown,
   L is the length of the rod, and θ is the angle the rod makes with the
   vertical (all angles here are measured in radians).

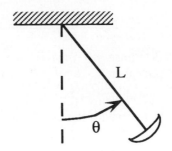

The maximum angle the rod makes is denoted as θ_M (so $-\theta_M \leq \theta \leq \theta_M$). The equation describing the motion of the pendulum comes from Newton's second law. One finds from this equation that the period of oscillation, T, is

$$T = 4\sqrt{\frac{L}{g}}\int_0^{\pi/2} \frac{d\phi}{\sqrt{1 - k^2 \sin^2\phi}} \quad ,$$

where

$$k = \sin(\tfrac{1}{2}\theta_M) \;,$$

and $g = 32$ ft/ sec^2 is the gravitational constant.

The pendulum we are going to consider comes from the short story by Edgar Allan Poe entitled "The Pit and the Pendulum." As you may recall, an unfortunate fellow is strapped to a low wooden table and is made to watch a pendulum swing back and forth above him. In addition to being tied down, his other problem is that the pendulum is descending (in his words, "down—still unceasingly—still inevitably down!"). He clearly understands the situation but he thinks that if he can use the blade at the end of the pendulum to cut his bonds, he will be able to jump out of the way before the pendulum returns for the "final" pass (actually he has a lot of help from some rats, but we'll not include them in our calculations). What you are asked to determine is how long he has to get off the table once the bonds are cut, that is, you are to calculate $T/2$. Keep in mind that this is a fellow who needs the calculations fast and he needs them accurate.

It appears from the narrator's description of his surroundings that the pendulum is about 32 feet long and makes a "terrifically wide" sweep, side to side, of 30 feet.

(a) What is θ_M ?

(b) Use the ideas described in Problem 1, parts (b) and (c), to find M_2 and M_4 .

(c) Use (3) to determine how large n should be to guarantee that the error in the trapezoidal approximation of T is no more than 10^{-6}.

(d) Use (4) to determine how large n should be to guarantee that the error in the Simpson approximation of T is no more than 10^{-6}.

(e) Which method of approximation should he use? Using this approximation, how much time does he have to get off the table?

17

Inverse Functions and Solving Transcendental Equations

Objective
To use Maple to solve transcendental equations by constructing an approximation of the solution and its inverse.

This Laboratory employs Maple to help demonstrate the exponential and logarithmic functions and the relationship between the two. In particular, we find that the concept of an inverse function arises quite naturally from our attempt to solve a deceptively simple equation involving the exponential. Our vehicle for this discussion is a mathematical model of the semiconductor diode—an electronic device of considerable past, present and future interest.

For a diode with a resistor in series, the current I flowing through the diode is related to the voltage V through the equation

$$I = I_s[\, e^{q(V - RI)/kT} - 1\,] \; . \tag{1}$$

There are five constants in this equation, and these are I_s, q, R, k, and T. We can eliminate them by changing variables to $i = I/I_s$, $v = qV/kT$, and $r = qRI_s/kT$. In this case we end up with the equation

$$i = e^{v - ir} - 1 \; . \tag{2}$$

Our objective in this Lab is to solve this equation for the current i and express it as a function of the potential v and resistance r . If you look at the equation for a moment you will see how difficult it is to solve for i , since it appears both on the left-hand side as well as in the exponential on the right-hand side.

EXAMPLE

To illustrate some of the ideas developed in the Laboratory problem suppose we want to solve the following equation for y :

$$y^3 = -\ln(x + y + 1) \ . \qquad\qquad (3)$$

Looking at this equation for a moment or two, one comes to the conclusion that there is no obvious solution for y . So, we can ask Maple to try to solve it:

```
> solve(y^3 = - ln(x + y + 1), y);
>
```

The second prompt is given here to emphasize the fact that Maple doesn't have a solution and therefore gives no response.

The next thing to try is plotting the solution. However, Maple needs explicit formulas to be able to construct a plot. To illustrate this suppose we try the following:

```
> plot(y^3 = -ln(x + y + 1), x = -1..1);
```

Error, (in plot) relational plots not implemented yet, y**3 = -ln(x+y+1)

This means that we are going to have to be a little more creative to be able to plot y as a function of x . Our approach will be to specify a particular value for x and then use `fsolve` to find the corresponding value of y . Interestingly enough, there are a couple of ways this can be done.

The most direct method uses `fsolve` and the arrow notation to define y = f(x) . This is done as follows:

```
> f := x -> fsolve( y^3 = - ln(x + y + 1), y);
```

It is now an easy matter to plot the function

```
> plot( f, x = -1..1);
```

Note that we did not use f(x) in the plot command, even though the arrow notation was used to define f .

To describe a second way to graph y as a function of x , suppose we want to graph y for $-1 \leq x \leq 1$. We now use `fsolve` to calculate y for many different values of x from this interval. For this example, we solve for y when $x = -1$, -0.9 , -0.8 , ..., 0.8 , 0.9 , 1 . To get Maple to do this repeated calculation we use a do-loop and the commands (along with comments explaining the commands) are

```
> eq := y^3 = - ln(x + y + 1);        # eq  designates the equation to be solved.
> for k from 0 to 20 do               # This starts the loop (k = 0 , 1 , 2 , ..., 20).
>    xpts[k] := -1+ k/10:              # These are the points along the x-axis.
>    EQ := subs( x = xpts[k], eq ):    # This evaluates the equation at  x = xpts[k].
>    ypts[k] := fsolve( EQ, y):        # This solves  EQ  for  y = ypts[k].
> od:                                  # A "reverse do"  designates end of loop.
```

When Maple finishes with this procedure, we will have 21 points, $k = 0..20$, where ypts[k] is the solution of (3) when $x = xpts[k]$. We now put these points together into a list of x,y-coordinates using the command

```
> points := [ [ xpts[j], ypts[j] ] $ j = 0..20 ];
```

To plot these, enter the following:

```
> plot( points, x = -1..1 , title = `Solution of Eq. (3)`);
```

Now that $y = f(x)$ has been determined we can also use the data points calculated from the do-loop to plot the inverse function $x = f^{-1}(y)$ (note this function is defined since the preceding plot shows $y = f(x)$ is monotone decreasing). To construct the y,x-coordinates we use the command

```
> ipoints := [ [ ypts[j], xpts[j] ] $ j = 0..20 ];
```

We plot these, together with "points," by entering

```
> plot({ipoints, points}, -1.1..1.1 , -1.0..1.3, title =`Plot of Function and Its Inverse`);
```

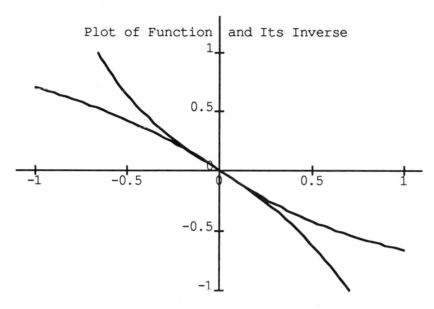

As an alternative to using points found from the do-loop, we can actually solve
(3) for x with the result that

$$x = -y - 1 + e^{-y^3}. \tag{4}$$

To compare this exact result with the numerically determined data in "ipoints" we
plot both on the same graph as follows:

> plot({ [y , −y − 1 + exp(−y^3), y = −1..1], ipoints }, title = `Comparison`);

It is apparent from this graph that there is little difference between the data and the
exact formula.

    You are asked to carry out similar calculations in the Laboratory. One
difference, however, is that the equation you are to solve, Eq. (2), contains a
parameter r . This generalization makes the problem more interesting. How one
deals with it is discussed in the Laboratory.

    Caution: In your Lab do not use the single capital letter I as a variable. This is
reserved for the imaginary number $\sqrt{-1}$ (however, you can use variables like I0
or II or i).

Laboratory Problems

The objective of the following problems is to investigate the solution of Eq. (2). In Problem 1 the case $r = 0$ is investigated, and in Problem 2 the case $r \neq 0$ is considered. The results of Problem 1 are used in Problem 2. In Problem 3 the exact solution is discussed and for this the results from Problem 2 are needed.

1. We begin by considering the case of zero resistance (i.e., $r = 0$) in Eq. (2).

   (a) Let izero be the current when $r = 0$ (note izero is a function of v).

   (b) Solve (2) for the potential v (when $r = 0$). This solution will be denoted by vzero (note vzero is a function of i).

   (c) Plot vzero and izero on the same graph using the interval $(-5, 5)$ for both axes. This is not as easy as it sounds, since izero is a function of v while vzero is a function of i . You will need to get around this somehow. One possibility is to use a command of the form

   > "plot({[x, f, x=a..b], [x, g, x=a..b]}, ...);"

   Explain how your plot shows that vzero is the inverse function of izero (and vice versa).

2. We now consider the more difficult case of nonzero resistance. We would like to solve (2) for i when $r \neq 0$. This can be done using a do-loop as in the example. The difference is that we will have to specify a value for r , and in the following a couple of values are considered.

   (a) Set $r = 1/2$. To solve (2) for i , using values of v from $v = -5$ to $v = 5$, use a do-loop as follows:

   ```
   > eq05 := subs( r = 1/2, i = exp(v − i*r) − 1 );
   > for k from 0 to 30 do
   >     v05[k] := −5 + k/3:
   >     EQ := subs( v = v05[k], eq05 ):
   >     i05[k] := fsolve( EQ, i ):
   > od:
   ```

What you should obtain from this loop is 31 points, where i05[k] is the solution of (2) when v = v05[k] and r = 1/2 . Put these points together using the command:

> s05 := [[v05[j], i05[j]]] $ j = 0..30];

Remember that if you want to suppress Maple's response you can terminate the command with a colon rather than a semicolon.

(b) Repeat the preceding calculations for r = 1 . In this way, construct the set of points s10 .

(c) Solve (2) for the voltage v and denote this solution as vi . In the case when r = 1 plot vi and s10 on the same graph using the interval (−5 , 5) for both axes. Is it clear from the plot why these functions are inverses of each other? Explain.

(d) Plot izero , s05 , and s10 on the same graph, using the interval [−5 , 5] for the horizontal axis. These represent the current-voltage characteristics for three resistance values. What effect does increasing resistance have on diode performance? For what voltage range does the effect of resistance appear to be most important?

3. (a) Using the points in the sets s05 and s10 , we have obtained approximations to the solutions of (2) for r = 1/2 and 1 . There is another way to do this using Maple. Use the `solve` command on Eq. (2) with nonzero resistance. You will find that Maple gives a reply in terms of an obscure function W . Find out about W by using the "?" (or "help") facility. Label Maple's answer for the current by the symbol iw .

(b) How do Maple's values for iw compare with the plots of the sets s05 and s10 ? Compare s10 and iw for r = 1 by plotting them together. What do you conclude?

18

The Trajectories of a Baseball

Objective
To investigate some questions about the path of a batted (or thrown) baseball, or other solid object, using parametric equations of motion.

Suppose that a baseball leaves the bat with an initial velocity of u feet per second at an angle A with the horizontal. We wish to determine the path followed by the ball.

First we consider a simplified version of the problem in which air resistance is neglected. Let v and w be the velocity components in the horizontal (x) and vertical (y) directions, respectively. Then, as a consequence of Newton's law, v and w satisfy the equations of motion

$$\frac{dv}{dt} = 0 \; , \quad \frac{dw}{dt} = -g \; , \tag{1}$$

where g is the acceleration due to gravity. The most general functions satisfying these equations are

$$v = c1 \; , \quad w = -gt + c2 \; , \tag{2}$$

where $c1$ and $c2$ are arbitrary constants of integration. To determine these constants we use the fact that the initial velocity is u. Splitting this into vertical and horizontal components, we end up with the conditions that $v(0) = u \cos(A)$ and $w(0) = u \sin(A)$. It follows that

$$c1 = u \cos(A) , \quad c2 = u \sin(A) ; \tag{3}$$

consequently,

$$v = u \cos(A) , \quad w = -gt + u \sin(A) . \tag{4}$$

Equations (4) give the velocity components of the moving baseball as functions of time t.

To find the position coordinates of the ball at any time we need to observe that $dx/dt = v$ and $dy/dt = w$. Then, integrating once more, we obtain

$$x = u \cos(A) t + c3 , \quad y = -\frac{1}{2} gt^2 + u \sin(A) t + c4 . \tag{5}$$

To determine the new set of arbitrary constants $c3$ and $c4$, we need the initial values of x and y. We assume that the origin is at the center of home plate, and that the bat strikes the ball at a height h above home plate. In this case the initial conditions are

$$x(0) = 0 , \quad y(0) = h . \tag{6}$$

It follows that $c3 = 0$ and $c4 = h$, and finally that

$$x = u \cos(A) t , \quad y = -\frac{1}{2} gt^2 + u \sin(A) t + h . \tag{7}$$

Equations (7) give the position coordinates of the baseball at any time t. They constitute a set of parametric equations for the path of the ball.

QUESTION 1

What kind of a curved path does the ball follow?

Answer
Since one of Eqs. (7) is linear in t and the other is quadratic, the path is a parabola. A plot of the path for some typical parameter values is shown in Fig. 1. To confirm that the path is a parabola you can easily eliminate t between these two equations, obtaining y as a quadratic function of x.

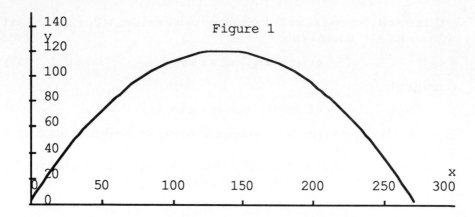

Figure 1

QUESTION 2

When does the ball reach the highest point on its trajectory? What is its maximum height?

Answer

Let us enter the expressions given by Eqs. (7) into Maple:

> x := u*cos(A)*t; y := −1/2*g*t^2 + u*sin(A)*t + h;

$$x := u \cos(A) t$$

$$y := - 1/2 g t^2 + u \sin(A) t + h$$

At the highest point on the trajectory the ball has ceased to rise and is on the verge of falling. Therefore dy/dt is zero. Thus

> yprime := diff(y, t);

$$yprime := - g t + u \sin(A)$$

> t1 := solve(yprime, t);

$$t1 := \frac{u \sin(A)}{g}$$

This is the time at which the ball reaches its highest point. The coordinates of the ball at this point are

> x1 := subs(t = t1, x); y1 := subs(t = t1, y);

$$x1 := \frac{u^2 \cos(A) \sin(A)}{g}$$

$$yl := 1/2 \; \frac{u^2 \, \sin(A)^2}{g} + h$$

The time t1 is also one-half of the time that the ball requires to return to its original height h . To confirm this

> t2 := solve(y = h, t);

$$t2 := 0, 2 \, \frac{u \, \sin(A)}{g}$$

Observe that the first value of t2 is the starting time, while the second is twice t1 . Also, observe that, for a given initial velocity u , both the times t1 and t2 , as well as the height y1 , are maximized when $\sin(A) = 1$, that is, when $A = \pi/2$. Thus the time in the air and the height that is reached are greatest when the batter pops the ball straight up.

The following problems extend the investigation of the flying baseball.

Laboratory Problems

In all of the following problems you may assume that $g = 32$ ft /sec^2 and that h = 3 ft.

1. Suppose that x and y are given by Eqs. (7):

$$x = u \, \cos(A) \, t \; , \quad y = -\frac{1}{2} \, g t^2 + u \, \sin(A) \, t + h \; .$$

   (a) If the outfield wall is a distance L from home plate, determine conditions on u and A so that a batted ball will leave the field of play (or hit the wall) while still rising.

   (b) If the batter is to accomplish this feat with the minimum expenditure of energy, that is, by imparting the minimum possible initial velocity to the ball, what should be the initial angle A ?

2. In Boston's Fenway Park the left field wall is approximately 38 feet high. It is 315 feet from home plate at the foul line and the distance increases as one moves toward center field.

   (a) Determine a relation between the initial velocity u and the initial angle A so that a batted ball will just barely clear the wall when it is a distance L from home plate.

(b) If $L = 340$ feet, plot the graph of u versus A from the relation found in part (a). Then determine the value of A for which u is minimum, and also find that minimum value of u .

The following problems deal with the effect of air resistance on the trajectory of the baseball and are somewhat more difficult.

3. Assume that the baseball in flight is subject to a force, due to air resistance, that is proportional to the velocity. Then, instead of Eqs. (1), the equations of motion are

$$\frac{dv}{dt} = -rv , \quad \frac{dw}{dt} = -rw - g , \tag{8}$$

where r is the coefficient of air resistance and the other letters have their previous meanings.

(a) Proceed, as in the discussion at the beginning of this Laboratory, to derive expressions for the position coordinates x , y of the ball at any time t . In other words, integrate Eqs. (8) twice, using also the additional conditions that

$$v(0) = u \cos(A) , \quad w(0) = u \sin(A) , \quad x(0) = 0 , \quad y(0) = h . \tag{9}$$

You should obtain

$$x = -\frac{\exp(-rt) \, u \cos(A)}{r} + \frac{u \cos(A)}{r} , \tag{10}$$

$$y = -\frac{gt}{r} - \frac{\exp(-rt) \, (u \sin(A) + g/r)}{r} + \frac{u \sin(A)}{r} + \frac{g}{r^2} + h . \tag{11}$$

(b) Find the limit as r approaches zero of the expressions found in part (a) for x and y . Are these limiting expressions the same as those in Eqs. (7)?

(c) Assuming that $u = 100$ ft/sec and that $A = \pi/4$ radians, plot the trajectory of the baseball for $r = 0.2$ and for $r = 0$. Describe the similarities and differences in the two trajectories.

(d) Assuming that $u = 100$, $A = \pi/4$, and $r = 0$ (no air resistance), determine the horizontal distance L that the ball travels before it hits the ground. (Assume that the ground is level.) Then, assuming the same value of A , but with a coefficient of resistance $r = 0.2$, determine the initial velocity u that is required in order for the ball to travel the same horizontal distance L .

4. Suppose that you are in Fenway Park with its left field wall approximately 38 feet high at a distance of 340 feet from home plate. Assuming that $r = 0.2$, determine the minimum initial velocity u that is required to clear the wall, and the corresponding initial angle A . Compare these results with those obtained earlier for the case of no air resistance.

Hint: You will probably have to use numerical and/or graphical means to solve this problem. For example, you might calculate u for several values of A , and then plot them, or interpolate in some way to estimate the minimum.

19

Logarithmic and Exponential Growth

Objective
To compare the growth of the logarithm and exponential functions with the power function.

   In this Laboratory you are to investigate the properties of the natural logarithm and exponential functions. Of particular interest is how fast they grow as $x \to \infty$ in comparison to the power function x^{α}. This is worth considering because there is a belief that exponential growth is exceedingly fast and logarithmic growth is exceedingly slow. For example, if given a choice between a job that offers an exponential increase in salary versus one that offers a logarithmic increase, few would hesitate on which choice to make. What we will find is that the answer is not so clear cut, and it is necessary to weigh other factors.

   In the first part of the Laboratory you are asked to explore the growth of these functions without using calculus. To explain the idea that is used, suppose we are given a function $y = f(x)$ and a particular point $x = x_0$. We can get a measure of the function's rate of growth by asking how far we must go along the x-axis, starting from x_0, to double the value of the function. For example, if $f(x) = x^2$ and we start at $x_0 = 2$, then we have to go to $x = 2^{3/2}$ to double the value of $f(2)$. For $g(x) = x^4$ we would have to go to $x = 2^{5/4}$ to double $g(2)$. Since $2^{5/4} < 2^{3/2}$ we conclude that $f(x)$ grows at a slower rate in this region than does $g(x)$. It is this sort of idea that is explored in the first problem of the Laboratory. In the second and third problems of the Lab you are asked to use calculus to explore the growth properties of the functions.

Laboratory Problems

In this Laboratory, Problems 2 and 3 are independent of one another but they both depend on Problem 1.

1. (a) To examine how fast $\ln(x)$ increases, determine how much x has to change to increase the value of $\ln(\cdot)$ by a factor of 2. In other words, if $\ln(x_0) = y_0$ then what value does k have to be so $\ln(kx_0) = 2y_0$? You can do this by hand or using Maple.

   (b) Determine k for the power function x^n. That is, find k so that if $x_0^n = y_0$ then $(kx_0)^n = 2y_0$. Assume here that $n = 10$. You can do this by hand or using Maple.

   (c) Determine k for the exponential function. That is, find k so that if $\exp(x_0) = y_0$ then $\exp(kx_0) = 2y_0$. You can do this by hand or using Maple.

   (d) Plot the values of the three k's you found in (a)–(c) on the same graph as a function of x_0 from $x_0 = 1$ to $x_0 = 10$. Based on these results comment on the growth of the natural logarithm, power, and exponential functions. Does your conclusion about which grows faster depend on x_0?

2. We used a fairly large exponent in the power function when comparing it with the logarithm so now we'll see how the $\ln(x)$ function compares with x^α when the exponent is fairly small. To do this let

$$f(x) = \frac{\ln(x)}{x^\alpha} \quad \text{for } x > 0.$$

The question we're going to consider is how does the growth of $\ln(x)$ compare with the growth of x^α when $0 < \alpha \le 1$?

   (a) Plot $f(x)$ when $\alpha = 1, 1/10, 1/100$ for $1 \le x \le 100$ (all three functions should appear on the same graph). Does it appear that $f(x)$ is increasing or decreasing? What does this mean about the growth of $\ln(x)$ in comparison to x^α?

(b)   Find the local maximum and minimum points for $f(x)$ when $0 < x < \infty$. Based on this determine in what intervals $f(x)$ is increasing and decreasing. Are your conclusions in (a) consistent with your answer here?

3.   In 1 (d) we investigated growth rates but said little about the actual values of the functions. In 2 (a) and (b) the logarithm was considered, so now we'll see how the exponential does against the power function. In a similar manner as before let

$$g(x) = \frac{e^x}{x^\alpha} \quad \text{for } x > 0 .$$

(a)   In the case when $\alpha = 10$, plot $g(x)$ for $1 \le x \le 10$. What conclusion about the values of the exponential function, in comparison with the values of the power function, can be made from this graph? Is your conclusion consistent with your conclusion in 1 (d)?

(b)   Find the local maximum and minimum points for $g(x)$ when $0 < x < \infty$. Based on this determine in what intervals $g(x)$ is increasing and decreasing.

It is common in applications to have to plot functions that range over several orders of magnitude. The problem with doing this is that a regular plot of such a function is not very informative. For example, the function $y = (x + 1)^4 \exp(-x^2)$ ranges between 1.48 to 1.8×10^{-8} for $1 \le x \le 5$, so when plotting the function most of it will appear to coincide with the x-axis. The traditional way to deal with this is to plot the function on what is known as semi-log paper (if you want to see what this stuff looks like, it is on display in most university bookstores). What one is doing when using a semi-log graph is plotting the common (base 10) logarithm of the function, that is, $\log(y)$ as a function of x. By doing this, $y = 10^{-2}$ corresponds to -2, $y = 10^{-8}$ corresponds to -8, etc. For the function $y(x)$ given earlier, the values along the vertical axis would range from about 1 down to -8. So, it's much easier to produce a graph that illustrates the behavior of the function. In Maple the common log function is $\log10(\cdot)$. Therefore to produce the same effect as semi-log paper in Maple one would use the plotting command

```
> y := exp(-x^2)(x + 1)^4:
```

```
> plot( log10(y) , x = 1..5 );
```

(c)   In the case when $\alpha = 10$ plot $\log10(g)$ for $1 \le x \le 40$. Is this graph consistent with your conclusion in 1 (d)? What about consistency with your conclusion in (b)? Make sure to comment on the interval used in 1 (d) and 3(a) compared with the interval used here.

By the way, you may want to make a regular plot of g(x) just to see how it compares with the log10(g) plot you obtained in (c).

20

Inverse Trigonometric Functions

Objective

To use inverse trigonometric functions to determine the maximum viewing angle in a theater.

In this Laboratory you are to determine the best place to sit in a movie theater. Which place is "best"? Well, the criterion used here is that the angle, θ, that is subtended by the eye when looking at the screen is as large as possible. A cross section through the middle of the theater is shown in Fig. 1. It consists of a wall (on the left) where the screen is located; the screen is H feet high and is h feet off the floor. There is a level stage between the wall and seating area that is L feet wide. The seating area is to the right of the stage, and the angle of inclination of the floor is α (remember that trigonometric functions in Maple require angles measured in radians). The distance up this incline that a person sits is denoted by x. We will assume the back wall of the theater is at $x = x_M$, so $0 \le x \le x_M$. In this Lab you are asked to find what value of x maximizes the angle θ.

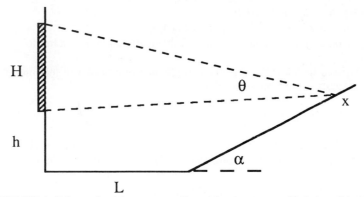

FIGURE 1 Schematic of the cross section of the theater used in Lab problem.

EXAMPLE

To illustrate how inverse trigonometric functions are used in this Lab, consider the case where there is no incline ($\alpha = 0$) and no stage ($L = 0$). This situation is shown in Fig. 2. In this case $\phi = \arctan(h/x)$ and $\theta + \phi = \arctan(\,(H+h)/x\,)$. Combining these equations we get

$$\theta = \arctan(\frac{H+h}{x}) - \arctan(\frac{h}{x})\ .$$

There are other ways to determine θ, and one is considered in Problem 1(b).

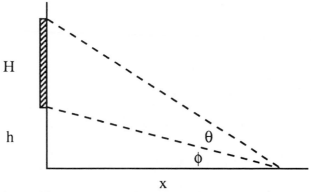

FIGURE 2 Schematic of the cross-section of the theater used in example.

Suppose for the theater that $H = 30$, $h = 10$, and $0 \le x \le x_M$, where $x_M = 40$. To see that indeed there is a value of x that maximizes θ, plot the function. This is done using the commands

> H := 30: h := 10: xM := 40;
> theta : = arctan((H + h)/x) − arctan(h/x);
> plot(theta, x = 0..xM, title = `Plot of Subtended Angle for Example`);

It is seen from the plot that there is a single maximum point in the interval. To find where it occurs we use the following:

> fsolve(diff(theta, x) = 0, x, 0..xM);

Maple responds with the answer 20 ; in other words, the best seat is in the middle of the theater. It is worth noting that in the `fsolve` command we restricted the search for solutions of $\theta'(x) = 0$ to the interval $0 < x < x_M$. This was done because this was the assumed size of the theater.

Laboratory Problems

As stated earlier we are going to consider the theater shown in Fig. 1. In the following problems you are asked to determine the best place to sit in this theater using the idea of maximizing the angle θ . In Problem 1 the general formula is discussed. In Problems 2 and 3, which are independent of one another, you are asked to apply this formula. Note that Problem 3(c) requires a little more than the usual computing, and some of the ideas discussed in Lab 16 make this problem easier to do.

1.   (a) Find θ in terms of x . Do this with the help of Fig. 3, by showing

$$\theta = \arccos\left(\frac{r^2 + s^2 - H^2}{2rs} \right) \, ,$$

where r and s are found using the formulas

$$r = [(L + x \cos \alpha)^2 + (h + H - x \sin \alpha)^2]^{1/2} \, ,$$

$$s = [(L + x \cos \alpha)^2 + (h - x \sin \alpha)^2]^{1/2} \, .$$

Hint: To derive the formulas for r and s draw horizontal lines back, from the top and bottom of the screen, to the vertical line that passes through the intersection point of the dashed lines. You can then get the formulas using a little trigonometry.

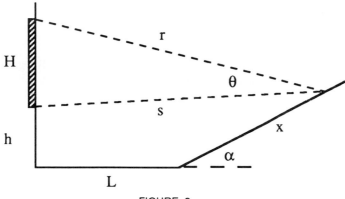

FIGURE 3.

(b) Show that the formula for θ in (a) also applies to the theater in Fig. 2, but with different formulas for r and s. Make sure to give these formulas.

(c) By examining the behavior of the argument of the inverse cosine function in (a), show that $\theta \to 0$ as $x \to \infty$. Explain what this limiting behavior means.

2. (a) For a local first-run movie theater $\alpha = 20°$, $L = 20$, $h = 10$, $H = 30$, and $x_M = 40$ (distances here are measured in feet). Plot θ as a function of x (for $0 \le x \le x_M$). Also, determine what value of x maximizes the angle θ. Assuming the rows are 3 feet apart and the first row is at $x = 0$, in which row should you sit?

(b) The IMAX® Theatre at the Kennedy Space Center is an impressive theater for movies of the space program. In this case, $\alpha = 17.7°$, $L = 28$, $h = 1/2$, $H = 52$, and $x_M = 44$. Redo (a) using these new parameter values.

(c) Comment on the differences between the theaters in (a) and (b). For example, does θ vary more in one than the other, or is θ much larger in one than the other? In terms of theater design, which do you consider more important, the minimum value of θ or the difference between the max and min values of θ ?

3. This part of the Lab concerns an auditorium at RPI where movies are shown. It is found that $H = 20$, $h = 10$, $L = 25$, and $\alpha = 20°$. Also $0 \le x \le x_M$, where $x_M = 45$. By the way, you may want to use values for a theater you are familiar with. This will require you to go to the theater, as the students at RPI are required to do, and measure (or estimate) the correct parameter values.

(a) Plot θ as a function of x .

(b) What value of x maximizes the angle θ ? What row does this correspond to?

(c) Suppose the theater owner wants to try to adjust things to have the best view possible throughout the entire theater, and it is decided to do this by making the average angle as large as possible. This average is given as

$$AVG := \frac{1}{x_M} \int_0^{x_M} \theta \, dx \ .$$

Assume the only parameter that can be adjusted is the height of the screen H . Moreover, the theater is constructed in such a manner that $H + h = 30$, where 30 is the height of the ceiling. So, once H is specified, then $h = 30 - H$. Plot AVG for $0 \le H \le 30$. From this graph estimate the value of H that maximizes AVG . [In answering this question you may find that Maple can't evaluate the integral quickly. If so, you might want to consider using Simpson's rule to obtain the value. This command is found in the `student` package and the use of this command is explained in the example of Laboratory 16.]

21

Applications of l'Hospital's Rule

Objective
To use l'Hospital's rule to find an approximation to an interesting but complicated function.

    In this Laboratory you are to investigate how l'Hospital's rule can be used to determine an approximation to a function. The problem to be solved has to do with the period of oscillation of a simple pendulum and how it depends on the maximum displacement of the weight. In introductory physics courses the period is calculated for the case when the weight does not move very far from its rest position. Here we will use Maple to find a more accurate approximation and then use this information to answer a particular question. It is worth pointing out that the method we will employ to obtain the approximation is very powerful and is one of the central components of what are known as asymptotic approximations.

    To describe the problem to be solved, suppose we are building a grandfather clock that has a pendulum with a rod that is 4 feet in length. The period of oscillation for the pendulum is supposed to be 2.25 seconds. The question is, what is the minimum width of the clock so this is possible? In our analysis the dimensions of the weight will be ignored.

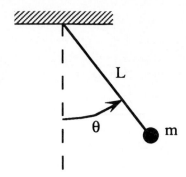

Before answering the question we first summarize a few facts about pendulums (see the figure). As shown, L is the length of the rod, m is the mass at the end, and θ is the angle the rod makes with the vertical (all angles here are measured in radians). The maximum angle the rod makes is denoted as θ_M (so $-\theta_M \le \theta \le \theta_M$). The equation describing the motion of the pendulum comes from Newton's second law. One finds from this equation that the period of oscillation T is

$$T = 4 \sqrt{\frac{L}{g}} \int_0^{\pi/2} \frac{d\phi}{\sqrt{1 - k^2 \sin^2\phi}} , \tag{1}$$

where

$$k = \sin(\tfrac{1}{2}\theta_M) ,$$

and $g = 32$ ft /sec^2 is the gravitational constant.

With the given integral, we can restate the earlier question. Namely, we are given T and L, and from (1) we want to determine θ_M. With this information we will then be able to find the minimum width of the clock using a little trigonometry. However, finding θ_M is not easy, and the best we can hope for is an approximation of its value. How to do this is an objective of this Laboratory.

In introductory physics it is usually assumed that θ_M is small enough that the period T can be approximated by its value when $\theta_M = 0$. In this case, from (1), one finds that $T \approx T_0$ where $T_0 = 2\pi \sqrt{L/g}$. The error made in using this approximation is small if θ_M is small. However, a better approximation is the following:

$$T \approx T_0 + T_1 (\theta_M)^\alpha , \tag{2}$$

where T_1 and α are constants. A way to determine these constants is outlined in the Laboratory problems.

For our clock, where $L = 4$, $T_0 \approx 2.22$ sec. This is close, but we want to do better. In particular, we expect that as θ_M increases that the period increases (this will be verified in the Lab problems), and we want to find just how big θ_M must be to get $T = 2.25$ sec. To do this we will use the approximation in (2).

EXAMPLE

To motivate how we will determine the constants in (2), recall the equation of the tangent line approximation to a function $f(x)$ at $x = x_0$:

$$f(x) \approx f(x_0) + f'(x_0)(x - x_0) \ . \tag{3}$$

Suppose we did **not** know the coefficients in this linear approximation to $f(x)$ and started out by assuming an approximation of the form

$$f(x) \approx f_0 + f_1(x - x_0) \ , \tag{4}$$

where f_0 and f_1 are unknown. Since this is to serve as an approximation for x near x_0, we will require it to be exact at $x = x_0$. So, letting $x = x_0$ in (4) we get $f_0 = f(x_0)$. To determine f_1 we can rewrite (4) as

$$f_1 \approx \frac{f(x) - f_0}{x - x_0} \ .$$

Again, we require the approximation to be exact as $x \to x_0$. However, we cannot simply set $x = x_0$ as before because the ratio on the right-hand side leads to an indeterminate form of type $0/0$. This is a situation where l'Hospital's rule can be used, and we can have Maple do this for us. Once Maple knows f and x0 we then use the commands:

```
> ratio := (f – f0)/(x – x0);
> f1 := limit( ratio, x = x0 );
```

Maple responds with the statement that $f_1 = f'(x_0)$. Therefore we have found that if we start out with the linear approximation in (4), we end up with the tangent line approximation given in (3). The approximation in (2) for T is a simple generalization of this and allows for the possibility that the best approximation may not be linear but some power function. In fact in the Laboratory that follows you will find that $\alpha \neq 1$.

Laboratory Problems

This Lab investigates a method for constructing an approximation to a function. Problem 1 uses the approximation in (2), while Problem 2 uses its generalization. For this reason Problem 2 depends on Problem 1.

1. (a) To get an idea of how T depends on θ_M, plot T for $0 \le \theta_M \le \pi/4$. To do this first determine 9 points on the curve for this range of θ_M's using a command such as the following (where thetaM = θ_M):

    > pts := evalf([[i*Pi/32, subs(thetaM = i*Pi/32, T)] $ i = 0..8]);

    Now plot these points using the command

    > plot(pts, title = `Plot of T vs. thetaM`);

    Explain why it does not appear that a linear approximation of T is worthwhile. Also comment on just how small θ_M must be for the introductory physics approximation to be accurate.

   (b) Solve (2) for α to obtain a solution of the form $\alpha \approx f(\theta_M)/g(\theta_M)$. Explain why

$$\lim_{\theta_M \to 0} \frac{f(\theta_M)}{g(\theta_M)}$$

   is an indeterminate form of type ∞/∞. Also explain why l'Hospital's rule can be used to determine the limit. Using (1) to determine T, calculate the limit and use this as the value for α.

   (c) Now that α is known, solve (2) for T_1 to obtain an expression of the form $T_1 \approx F(\theta_M)/G(\theta_M)$. Explain why

$$\lim_{\theta_M \to 0} \frac{F(\theta_M)}{G(\theta_M)}$$

   is an indeterminate form of type $0/0$. Also explain why l'Hospital's rule can be used to determine the limit. Calculate the limit and use this as the value for T_1.

   (d) At this point we should be concerned about the accuracy of our approximation. Address this by plotting, on the same graph, pts and the approximation in (2), making use of your values from (b) and (c). Comment on the accuracy.

(e) With the constants in (2) now determined answer the original question; that is, find the minimum width of the clock. Do you feel, based on your value for θ_M and the graph in (d), that this is an accurate result?

2. (a) If (2) is taken a step further, one gets the approximation

$$T \approx T_0 + T_1(\theta_M)^\alpha + T_2(\theta_M)^\beta \ ,$$

where T_0, T_1, α are as before, and T_2 and β are constants to be determined. Find these constants using the ideas developed in 1 (b) and (c).

(b) How much does the minimum width of the clock change using this more accurate approximation? Comment on this change and the accuracy of the approximation shown in your plot from 1 (d).

Extensions of the Laboratory

The ideas developed in this Laboratory can be used on other problems. In the following, for each function $y(x)$, the form of the approximation for y when x is close to $x = x_0$ is given. Solve for the constants (as in 1 (b) and (c)), explain why l'Hospital's rule can be applied, and then determine the constants.

1. $y = \sin(x)$ where $x_0 = 0$ and $y \approx y_0 x^\alpha$. How does this approximation differ, if at all, from the tangent line to $y = \sin(x)$ at $x = 0$?

2. $y = \exp(\cos(x))$ where $x_0 = 0$ and $y \approx y_0 + y_1 x^\alpha$. How does this approximation differ, if at all, from the tangent line to $y = \exp(\cos(x))$ at $x = 0$?

3. $y = \displaystyle\int_0^x \exp(-s^2)\, ds$ where $x_0 = 1$ and $y \approx y_0 + y_1(x - 1)^\alpha$.

4. $y = \dfrac{1}{1 - \cos(x)}$ where $x_0 = 0$ and $y \approx y_0 x^\alpha$.

5. In (4), using your values of y_0 and α, use the approximation $y \approx y_0 x^\alpha + y_1 x^\beta$.

22

A Glimpse of Infinity

Objective
To investigate some commonly occurring improper integrals with the goal of
improving your understanding of the limiting behavior of functions at infinity.

One of the goals of this Lab is to improve your understanding of the behavior
of functions $f(x)$ as x approaches infinity. Another goal is to help you develop
some intuition regarding whether or not a specific improper integral converges or
diverges.

In order to evaluate the improper integral

$$\int_1^\infty f(x)\, dx \ ,$$

we first evaluate the integral

$$\int_1^b f(x)\, dx \ ,$$

and then we take the limit of the resulting expression as $b \to \infty$. If this limit is
finite, we say the integral converges; otherwise we say that the integral diverges.

EXAMPLE 1

Show that the integral int(1/x, x = 1..infinity) diverges.

First, we evaluate the integral over the interval $1 \le x \le b$:

> int(1/x , x = 1..b);

$$ln(b)$$

Of course, ln(b) approaches infinity as $b \to \infty$ and this is confirmed with the command:

> limit(ln(b), b = infinity);

$$infinity$$

Maple does this two step process automatically if we issue the command

> int(1/x, x = 1..infinity);

Thus, there is infinite area under the curve over the interval $[1, \infty)$, and so, we say that the integral diverges.

Now, you would think that since the area is infinite over the interval $[1, \infty)$ that if b is large, then the area over the interval $[1, b]$ is large. To investigate this we take $b = 10^{100}$. This is reasonably large, in fact, it is larger than the number of atoms in the universe. Therefore one might think that the area under 1/x for $1 \le x \le 10^{100}$ must be very large indeed. Input the Maple command to find this area as 230.258... . Thus, most of the area under the curve 1/x is from 10^{100} on out, even though the curve 1/x is within 10^{-100} of the x-axis when $x = 10^{100}$! In fact, no matter what large number we pick, there is an infinite area under the curve 1/x to the right of that number.

EXAMPLE 2

Show that the integral int(1/x**2, x = 1..infinity) converges.

Based on what occurred in the last example, one might be tempted to think that this integral also diverges.   However, the area under this curve for $1 \le x < \infty$ is actually equal to 1. Thus, the integral converges.
Solution Outline:

(i)   Input the Maple commands (as in Example 1) to evaluate the integral over the interval $1 \le x \le b$ and then take the limit as $b \to \infty$ to show that the area under the curve $1/x^2$ for $1 \le x < \infty$ is in fact equal to 1.

(ii)   You should realize that $1/x^2$ approaches the x-axis much faster than $1/x$ does. Plot both curves on the same axes over $[1, 20]$.

(iii)   Input the Maple command to find the area under $1/x^2$ for $1 \le x \le 10^{100}$. When Digits := 10, you should get 1.0000000000 when you use the command

$$\text{evalf(int(} 1/x**2, x = 1..10**100));$$

Set Digits := 100 and redo the calculation.

(iv)   Input the Maple command to find the area under $1/x^2$ for $10^{100} \le x < \infty$ when Digits: = 100. Be sure to reset Digits to 10 when you are done.

This illustrates that there is only a tiny amount of area under the curve $1/x^2$ to the right of any very large number.

EXAMPLE 3

Show that the integral int($1/x**p$, x = 1..infinity) converges provided that $p > 1$.

In the previous two examples, we saw that the area under $1/x$ for $1 \le x < \infty$ is infinite but the area under $1/x^2$ is finite. In this example, we investigate what happens for functions of the form $1/x^p$ where $p > 1$.
    If $p = 1.1$, we first evaluate the integral over $1 \le x \le b$:

> int(1/x**1.1, x = 1..b);

$$-\frac{10}{b^{1/10}} + 10.$$

As $b \to \infty$, this expression approaches 10 as is confirmed by the following:

> limit(" , b = infinity);
$$10.$$

We could have used the single command

> int(1/x**1.1, x = 1..infinity);

Of course, it is easy to do the preceding integral by hand. You should work out this integral with pencil and paper just to make sure you can do it.
    For $p = 1.01$ we get

> int(1/x**1.01, x = 1..infinity);
$$100.$$

For $p = 1.001$, we get

> int(1/x**1.001, x = 1..infinity);
$$1000.$$

Thus a reasonable (and correct) guess is that

$$\text{int(} 1/x^{**}p \text{ , } x = 1..\text{infinity}) = \frac{1}{p-1} \quad \text{provided } p > 1 \text{ .}$$

Therefore no matter how close p gets to 1 , the area under the curve $1/x^p$ is finite provided $p > 1$. However, the area under the curve $1/x$ is infinite. As additional exercises you might try the following:

(i)   Find the area under $1/x^{**}1.001$ over the interval $[1 , 10^{**}100]$. You should get about 205.6 .

(ii) Find the area under $1/x^{**}1.001$ over the interval $[10^{**}100 , \infty)$. You should get about 794.4 .

Thus, the area under the curve $1/x^{**}1.001$ for $x > 10^{**}100$ is finite, but the area under the curve $1/x$ for $x > 10^{**}100$ is infinite.

Laboratory Problems

The two problems below apply the ideas developed in the examples. Problem 1 is similar to the example problems, and Problem 2 extends the approach.

1.   In this problem, you are asked to investigate the improper integral

$$\text{int(} 1/(x^*\ln(x)), x = 2..\text{infinity}) \text{ ,}$$

and explore the relationship of the function $1/(x^*\ln(x))$ with the functions discussed in the three examples. Notice that the integral is from $x = 2$ to $x = \infty$ because $1/(x^*\ln(x))$ is not defined when $x = 1$.

(a) Use Maple to evaluate int($1/(x^*\ln(x)), x = 2..b)$.

(b) Use the method of substitution to evaluate this same integral by hand. You (and Maple) should have obtained $\ln(\ln(b)) - \ln(\ln(2))$. As $b \to \infty$, this expression clearly approaches ∞ . Thus, the area under the curve $1/(x^*\ln(x))$ over $[2 , \infty)$ is infinite. The single Maple command

$$\text{int(} 1/(x^*\ln(x)), x = 2..\text{infinity});$$

will also yield this same result. Issue this command to check that this is true.

(c) Plot the two functions $1/(x*\ln(x))$, and $1/x**1.001$ over the interval $[2, 10]$. From this plot we see that for $3 < x < 10$, the curve $1/(x*\ln(x))$ is below $1/x**1.001$. It cannot be true that the curve $1/(x*\ln(x))$ remains below the curve $1/x**1.001$ as $x \to \infty$. Explain why, using what you know about the areas under these two curves.

(d) The curve $1/(x*\ln(x))$ must eventually intersect and remain above the curve $1/x**1.001$. The value of x at which this occurs is amazingly large. For example, if we evaluate both functions at $x = 10**100$, we obtain

> a := evalf(subs(x = 10**100, 1/x**1.001));

$$a := .7943282347*10^{-100}$$

> b := evalf(subs(x = 10**100, 1/(x*\ln(x))));

$$b := .4342944819*10^{-102}$$

Thus even at $x = 10**100$, the curve $1/(x*\ln(x))$ is still below the curve $1/x**1.001$, because $b < a$.

Use Maple with the `evalf` and `subs` commands that have been illustrated to find the first power of 10 where the curve $1/(x*\ln(x))$ is finally above $1/x**1.001$. *Hint:* Look in the range from $10**3955$ to $10**3965$.

(e) Even though the area under the curve $1/(x*\ln(x))$ over $[2, \infty)$ is infinite, there is a surprisingly small area under this curve over the interval $[2, r]$ where r is a very large number. Use Maple to show that the area under $1/(x*\ln(x))$ for $2 \le x \le 10**3960$ is only about 9.5. *Caution:* Use `evalf(int(....));` otherwise you will get 3960 zeros printed.

Remember that the area under this curve from $x = 10**3960$ to ∞ is infinite. Thus for some improper integrals it is extremely important to examine what happens for large values of x, and for some functions amazingly large values need to be considered before one can get a sense of what it happening.

(f) In part (d), we saw that for $p > 1$, $1/(x*\ln(x)) > 1/x**p$ for large values of x. Letting $p = 1 + q$, we see that for $q > 0$, $1/(x*\ln(x)) > 1/x**(1+q)$ for large values of x. With paper and pencil show that for any $q > 0$, $\ln(x) < x**q$ for large values of x. This shows that, as $x \to \infty$, $\ln(x)$ grows more slowly than x^q for any positive value of q.

2. In the examples, we saw that int(1/x, x = 1..infinity) diverges but the integral int(1/x**p, x = 1 .. infinity) converges when $p > 1$. Consider the function $1/x$ over the interval $[1, \infty)$. Suppose that we rotate this

curve about the x-axis to obtain a three-dimensional object that looks like a horn.

(a) The outside (and the inside) surface area of this horn is given by the limit of the following integral as $b \to \infty$:

surface area := int(2*Pi*(1/x)*sqrt(1 + 1/x**4) , x = 1..b)

If you forgot the formula for surface area, look it up and check that the preceding is correct. Maple cannot evaluate this integral symbolically, but for specific values of b it can perform a numerical integration. Use Maple to show that for b = 10**100 , the surface area is about .1375505554*10^{99} .

Actually, because sqrt(1+1/x**4) > 1 we see that

$$(1/x)*sqrt(1+1/x**4) > 1/x ,$$

and so

$$2*Pi*(1/x)*sqrt(1+1/x**4) > 2*Pi*1/x .$$

But the integral, int(2*Pi*1/x, x = 1..infinity) , diverges so the preceding integral for surface area must also diverge.

(b) Show that the volume of the three-dimensional region inside the horn is equal to π .

(c) Wait a minute, how can this be? The surface area is infinite and the volume is finite! Suppose that you wanted to paint the outside of the horn with a paint that covered 2400 sq ft per cubic foot of paint. This would be the coverage you would get if the paint were put on with a thickness of 1/200 of an inch. Our calculation in part (a) shows that it would take an infinite number of cubic feet of paint to cover the outside, but to fill up the inside would only take about 3.1416 cubic feet of paint. If we fill up the horn with paint, surely the inside surface area is covered with paint. And the inside surface area equals the outside surface area. This seems like a contradiction! Explain why this is not a contradiction. It will help you to think about how much paint it would take to paint the inside surface area and about how our assumption on the thickness of the layer of paint enters into this calculation.

23

Understanding Convergence of Sequences and an Example of Chaos

Objective

To calculate and plot the terms in a sequence as a means of understanding how it behaves, and an example of chaos.

This Laboratory has to do with sequences, which are simply lists of numbers denoted by $a(1)$, $a(2)$, ..., $a(n)$, ... We usually are interested mainly in the behavior of a sequence for large values of n. For example, we want to know whether or not the sequence converges to a single number. Or does it oscillate among two or more numbers? Or does it become unbounded? Or what?

In this Lab you will learn how to calculate and plot the terms in a sequence as a means of understanding how it behaves. Moreover, by working out the second problem that follows, you will get a glimpse of chaos, currently a very active topic of investigation in many technical fields.

EXAMPLE 1

Many sequences are defined explicitly, that is, there is a formula for the general term $a(n)$. For example,

> a := n -> 1/(n^(3/2) + 4);

$$a := n \;\rightarrow\; \frac{1}{n^{3/2} + 4}$$

After this, we will omit Maple's response. To see it, you should issue the command. To obtain the value of a particular term of the sequence, such as the tenth, use one of the following commands:

```
> a(10);
> evalf( a(10) );
```

To see several terms in this sequence, use a $ to tell Maple to repeat the command as follows:

```
> a(n) $ n = 1..10;
```

If you prefer decimal numbers invoke the `evalf` command as follows (note the 1.0 and 10.0);

```
> evalf( a(n) ) $ n = 1.0..10.0;
```

To plot the terms in a sequence, we must first construct a list containing the coordinate pairs. Each pair [n , a(n)] contains the index and the corresponding term of the sequence. For instance, to plot the first 10 terms we construct the list

```
> A := [ [ n, a(n) ] $n = 1..10];
```

Observe the use of the square brackets in the preceding command. This list of points can now be plotted using the plot command. You can obtain either a point plot or a line plot by inserting the option "style = POINT" or "style = LINE" in the plot command. The point plot shows the actual terms in the sequence, but the line plot is sometimes easier to read. You should try them both:

```
> plot( A, style = POINT );
> plot( A, style = LINE );
```

It appears from the plot, or from an inspection of the expression for a(n) , that this sequence has the limit zero as $n \rightarrow \infty$.

To find how far out in the sequence we must go in order to make sure that terms in the sequence are within a given distance of the limiting value, we can use the `fsolve` command. For example,

```
> fsolve( a(n) – 0 = 0.01, n );
```

This shows us that we must go to the 21st term before a(n) is within 0.01 of its limiting value. Let us now try an arbitrary tolerance eps (for epsilon):

```
> solve( a(n) – 0 = eps, n );
```

Note that the last two values are complex, and hence irrelevant. Let us denote the first value by N :

```
> N := "[1];
> evalf(subs(eps = 0.01, N ));
```
$$20.96593115$$

This agrees with the value found previously.

EXAMPLE 2

Some sequences are defined recursively, that is, a(n) is expressed in terms of one or more of the earlier terms. For example, let us suppose that

$$a(n) \; = \; \frac{1}{1 + \text{sqrt}\,(a(n-1))} \;\; , \; n \, = \, 1\,,2\,,... \;\; .$$

Thus each term in the sequence is the reciprocal of the square root of the preceding term plus one. If we are given the value of a(0), then we can determine other terms in the sequence by using this formula over and over. This is known as iterating the formula.

The following Maple procedure is designed to accept a given function and a given initial value, and then to iterate the function n times. Type the following commands (following the >) exactly as they are given here:

```
> iterate := proc(f, a0, n) local i,j;
>     a(0) := evalf(a0);
>     for i from 1 to n do
>     a(i) := evalf( f(a(i−1)) )
>     od;
> a(j) $ j = 0..n;
> end;
```

To save this procedure so that you can use it again in a later session, you should now issue the following `save` command:

```
> save `iterate.m`;
```

When you wish to use this procedure again, you can now bring it back by issuing the command

```
> read `iterate.m`;
```

Now define the function f for the sequence given in the preceding:

```
> f := a −> 1/( 1 + sqrt(a) );
```

Now let us find the first 10 terms in this sequence, starting with a0 = 0.75 :

```
> iterate( f, 0.75, 10 );
```

Now repeat the last step for one or two other (nonnegative) initial values a0 . You might also want to plot the terms in this sequence as shown in Example 1.

Finally, as a word of caution when using the procedure `iterate.` Anytime you want to switch formulas for the function f(a) you should first enter the command:

> a := 'a':

and then enter the new formula for f(a) .

There are several questions we will address in the problems, and they include the following:

    (i)   Does the sequence have a limit L ?

    (ii)   Does the value of L depend on the starting point a0 ?

    (iii)   What is the limit L ?

Laboratory Problems

In Problem 1 you are to investigate a sequence defined explicitly and in Problem 2 you consider one defined recursively. The two problems are independent of one another.

1.   Consider the sequence

$$a_n = 2 + \frac{1}{n^2 + n} \ , \ n = 1, 2, \dots \ .$$

(a) Calculate and then plot the first 20 terms in this sequence. Does this sequence appear to have a limit L ? If so, what is L ?

(b) Determine how far in the sequence you must go so that the terms are within 0.001 of the limiting value L . Let N1 be the value of n for which a_n is first within 0.001 of L .

(c) Let ε be an arbitrary (small) positive number. Determine how far in the sequence you must go in order that the terms are within ε of the limiting value L . Let $N(\varepsilon)$ be this value of n .

(d) Substitute $\varepsilon = 0.001$ into $N(\varepsilon)$. Do you obtain the same value as the number N1 from part (b)? If you do not, explain why not.

(e) Describe how $N(\varepsilon)$ depends on ε for small positive values of ε .

2.   Consider the sequence defined by the recursive relation

$$a_{n+1} = ra_n(1 - a_n) \ .$$

This equation is known as the logistic difference equation. Our goal here is to explore how the terms a_n in the sequence behave, and how their behavior changes as the parameter r changes. You should use the procedure `iterate` defined in Example 2. Throughout this problem you should choose initial values in the interval $0 < a_0 < 1$.

(a) Let $r = 3/2$. Calculate a moderate number of terms in the sequence (perhaps 10 to 20). Does the sequence appear to be converging? What do you think its limit is? Does the limit depend on your choice of the initial value? Plot the terms in the sequence that you have calculated.

(b) Let $r = 2.8$. Again calculate a moderate number of terms in the sequence. Does the sequence appear to be converging? If so, to what value? Plot the terms that you have calculated. How does the behavior of the sequence in this case differ from that in part (a)?

(c) Let $r = 3.2$, and calculate some terms in the sequence. Show that the sequence does not appear to converge, but rather it behaves in a different way. Plot the terms that you have calculated, and describe in your own words how the sequence behaves in this case.

(d) You should have found that the sequence converges to a single value for $r = 2.8$ and that it approaches a steady oscillation between two values for $r = 3.2$. Consider intermediate values of r and try to determine more precisely where this transition, or bifurcation, takes place. For each case that you investigate determine enough terms so that it is clear to you what is happening. It is better to **plot** the terms rather than printing out the numerical values.

(e) Consider values of r in the range $3.43 < r < 3.46$. Note that for a certain value of r the sequence begins to oscillate among four values, rather than two. In other words, the period of the oscillation has doubled. Determine as accurately as you can the value of r for which this happens.

(f) As r continues to increase, further period doublings occur. Try to find the next one, that is, the value of r for which the sequence begins to oscillate among eight values.

(g) For r larger than about 3.57 the terms in the sequence appear to wander in a nonperiodic, indeed in a chaotic, manner. Choose such a value of r, for example, $r = 3.65$, and calculate a considerable number of terms (at least a few hundred). Again, it is better to **plot** the terms rather than printing out the numerical values. Observe that no matter how many terms you calculate, it is not possible to predict the values of the next few terms. Thus, the sequence is **unpredictable**.

(h) For $r = 3.65$ choose two different initial values, that is, values of a(0), that differ by 0.001. For example, you might choose 0.3 and 0.301. Calculate and/or plot some terms in the sequence for each initial value. Determine how long the terms in the two sequences remain close together, and when they begin to depart significantly from each other. This sequence **depends sensitively** on the initial condition.

Note: The peculiar behavior of this apparently simple sequence was discovered in about 1974 by the Australian mathematical biologist Robert May. It was one of the early examples of what has come to be known as chaos in mathematical systems, and it helped to stimulate an enormous amount of investigation of such phenomena in the last two decades. Much has been learned, but many mysteries remain.

24

Convergence of Infinite Series and the Ratio Test

Objective

To use Maple to investigate the convergence and sum of an infinite series, and to examine the reason why the ratio test works.

Of all the tests that have been devised for determining convergence of infinite series, the ratio test is probably the best known. However, it is likely that if you ask someone why the test works they won't be able to tell you even though they may use it on a regular basis. One of the objectives of this Laboratory is to investigate why it works. As we will see, it relies heavily on the geometric series and the fact that we know when a geometric series converges and when it doesn't.

Another objective of the Laboratory is to show how to use Maple to carry out the ratio test and then calculate the sum. As will be demonstrated in the following example, the `limit` command in Maple can be used to determine convergence or divergence very easily. However, like most everything else in life, Maple has bounds on what it can do. This will also be demonstrated.

EXAMPLE

To illustrate some of the Maple commands you will need in the Laboratory problems, consider the series

$$\sum_{k=1}^{\infty} \frac{k^4}{2^k} \ . \qquad\qquad (1)$$

We can get Maple to use the ratio test to determine whether the series converges by using the commands:

> a := k -> k^4/2^k;   # This defines the k-th term a_k of the series.
> L1 := limit(a(k+1)/a(k), k = infinity);   # This calculates the limit.

One finds that L1 = 1/2 , which means the series converges. It appears from this example that Maple can be very useful in applying the ratio test. You will find that it is capable of doing many, if not most, of the exercises on the ratio test in any calculus text. However, for some it may be necessary first to manipulate the ratio. An example of this arises with the series

$$\sum_{k=1}^{\infty} \frac{2(2k)!}{(k!)^2} \ . \qquad\qquad (2)$$

To deal with this one we include the `expand` command as follows (we use b_k to designate the k-th term for this series):

> b := k -> 2*(2*k)!/(k!)^2;
> ratio := expand(b (k+1)/b (k));
> L2 := limit(ratio , k = infinity);

One finds that L2 = 4 , and so the series diverges. For others it is not clear what steps are necessary to help Maple determine the limit. An example of such a situation is the series

$$\sum_{k=1}^{\infty} \frac{(2^k)!}{2^{k!}} \ . $$

You might try to determine on your own whether or not this series converges.
       To find the sum of the series in (1) we need to calculate the partial sums

$$s_n = \sum_{k=1}^{n} \frac{k^4}{2^k} \ . $$

To tell Maple about the function s(n) = s_n use the `sum` command as follows:

> s := n -> sum(a(k), k = 1..n);

The next objective is to plot the partial sums s(n) for n from 1 to 30. We go up to n = 30 only because this gives us a good idea of to what value the series converges. To construct the plot we first calculate the coordinates [[1 , s(1)] , [2 , s(2)] , ..., [30 , s(30)]] using the command

> points := evalf([[i, s(i)] $ i = 1..30]);

By looking at the values in points do you think it is reasonable to say that with s(30) we have the value of the sum of the series to at least five decimal places? A way to prove this is discussed in Laboratory 25.

  Now, to plot these points enter the following:

> plot(points, style = POINT, title = `Partial Sums for Example`);

Its important that the word POINT in this command be in capital letters. If you prefer to have the points connected by straight lines, replace POINT with LINE .

Laboratory Problems

For Problems 1 and 2, you are asked to investigate the series

$$\sum_{k=1}^{\infty} \frac{\ln(k + 1)}{(\ln(10))^k} . \tag{3}$$

Because of this, Problem 2 depends on Problem 1. In Problem 2 you are asked to investigate why the ratio test proves a series converges. The reason why the ratio test can be used to prove divergence is taken up in Problem 3, and for this you are asked about the series in (2). Problem 3 is independent of both Problems 1 and 2.

1. The following two parts are direct applications of Maple to the study of an infinite series.

   (a) Letting a_k be the k-th term in the series in (3), use Maple to calculate

$$L3 = \lim_{k \to \infty} \frac{a_{k+1}}{a_k} .$$

   Does the series converge?

   (b) Setting

$$s_n = \sum_{k=1}^{n} \frac{\ln(k+1)}{(\ln(10))^k} \, ,$$

plot s_n for $1 \le n \le 20$. Is the plot consistent with your conclusion in (a)? How many terms in the series does it appear that you need to find the sum to four-place accuracy?

In the problems to follow you are asked to investigate why the ratio test works. For example, in Problem 2 you begin with the observation that $L3 < 1$ and then deduce that the series in (3) converges. To do this you will compare it with the series

$$\sum_{k=1}^{\infty} r^k = \frac{r}{1-r} \, . \qquad (4)$$

This is just the geometric series, except that it starts at $k = 1$ rather than at $k = 0$. This series is chosen because we know exactly when it does, and when it doesn't, converge. Namely, it converges only when $-1 < r < 1$.

2. To investigate why the ratio test can be used to establish convergence, we are going to pick r so that it satisfies the inequalities $L3 < r < 1$. Any number in this interval will work for what we want to accomplish, but to be definite we will take $r = (L3 + 1)/2$, that is, the midpoint.

   (a) For the series in (3), by plotting a_{k+1}/a_k for $1 \le k \le 100$, explain why

$$a_{k+1} < r a_k \ \ (\text{at least for } 1 \le k \le 100 \ ).$$

   Since $a_1 = \ln(2)/\ln(10) < 1$, the result in (a) means that $a_k < r^k$. It's possible to prove this mathematically, but we prefer to show it graphically.

   (b) Letting $AR = a_k/r^k$, plot AR for $1 \le k \le 100$. Explain why this plot indicates that the comparison test is a good candidate to establish the convergence of the series in (3). Note that the plot cannot be used for a *proof* of convergence, since it doesn't include the interval $100 < k < \infty$.

   (c) To summarize what you did in (a) and (b), you used the fact that $L3 < 1$ to conclude that the series in (3) converges. It is this conclusion that is the basis of the ratio test. Based on these observations, does it seem fair to say that the ratio test for convergence is nothing more than a special case of the comparison test using the geometric series? If so, why does the condition $L3 < 1$ imply convergence in the ratio test?

3. Now that convergence is understood we turn our attention to why the ratio test can also be used to show divergence. The series we will examine is

given in (2). We already know $L2 > 1$, and so the question is, why does this mean the series diverges? As before we will compare the series with the geometric series in (4), but now r is to satisfy $1 < r < L2$. To be definite, we'll take $r = (L2 + 1)/2$.

(a) For the series in (2), by plotting a_{k+1}/a_k for $1 \leq k \leq 10$ explain why

$$a_{k+1} > ra_k \text{ (at least for } 1 \leq k \leq 10).$$

One consequence of the result in (a) is that $a_k > r^k$. It's possible to prove this mathematically but we prefer to show it graphically.

(b) Letting $BR = a_k/r^k$, plot BR for $1 \leq k \leq 10$. Explain why this plot indicates that the comparison test is a good candidate to establish the divergence of the series in (2). Note that the plot cannot be used for a *proof* of divergence since it doesn't include the interval $10 < k < \infty$.

(c) To summarize what you did in (a) and (b), you used the fact that $L2 > 1$ to conclude that the series in (2) diverges. It is this conclusion that is the basis of the ratio test. Based on these observations, does it seem fair to say that the ratio test for divergence is nothing more than a special case of the comparison test using the geometric series? If so, why does the condition $L2 > 1$ imply divergence in the ratio test?

25

Calculating the Sum of an Infinite Series

Objective

To calculate an accurate approximation of the sum of an infinite series using information derived from the integral test.

The sum of a convergent infinite series appears, at first glance, to be easy to calculate. One simply starts adding up the terms in the series, one after the other. This is relatively easy to do, but the problem is that there are an infinite number of terms. This usually means that the best we can hope for is to calculate an approximation of the sum. There is nothing wrong with this, since in principle we should be able to calculate the sum as accurately as we want. For example, if we want the sum to be correct to four decimal places we should be able to achieve this by simply adding up enough terms. The question is how many terms are needed to accomplish this task? Also if that number turns out to be very large, we would like to know if there is anything that can be done to speed up the process. One of the few methods that have been devised to answer these questions involves the integral test, and this Laboratory investigates how this is done.

To illustrate the ideas, suppose we are interested in finding the number s so that

$$s = \sum_{k=1}^{\infty} \frac{1}{2^k} .$$

(1)

173

With this in mind, suppose we add the first 10 terms of this series together to obtain the partial sum

$$s_{10} = \sum_{k=1}^{10} \frac{1}{2^k} . \tag{2}$$

The question is, how close is s_{10} to the sum s? This can be answered if we can determine the size of the remainder $R_{10} = s - s_{10}$.

The integral test gives us a way to estimate the remainder R_{10} and, consequently, the sum s. To explain how, it is worth reviewing the statement of the test. To do this, suppose we start out with the series

$$s = \sum_{k=1}^{\infty} a_k , \tag{3}$$

where $a_k \geq 0$. Also suppose we can find a function $f(x)$ that satisfies
   (i) $f(k) = a_k$ for $k = 1, 2, 3, ...$ and
   (ii) $f'(x) \leq 0$ for $x \geq 1$.
In this case, if s_n is the n-th partial sum and $R_n = s - s_n$, then $K_n - a_n \leq R_n \leq K_n$, where

$$K_n = \int_n^{\infty} f(x)\, dx . \tag{4}$$

Based on this theorem we get the following methods to approximate s:
(a) The standard approximation of the sum used in most calculus books is

$$s \approx s_n , \tag{5}$$

and the maximum error using this approximation is K_n.

(b) Another approximation can be obtained by using the midpoint between the upper and lower bounds on R_n given earlier. One finds that $R_n \approx K_n - a_n/2$, with a maximum error of $a_n/2$. Since $s = s_n + R_n$, we therefore obtain the following approximation for the sum:

$$s \approx s_n - \frac{1}{2} a_n + K_n , \tag{6}$$

with a maximum error of $a_n/2$.

In this Laboratory you are going to investigate the accuracy of these two approximations.

EXAMPLE

To incorporate Maple commands into our example let's start with the partial sum s_{10} given in (2). The command to calculate this is

> s10 := sum(1/2^k, k = 1..10);

Now to apply the integral test to the series in (1), we take $f(x) = 2^{-x}$. You should verify for yourself that this function satisfies the conditions of the integral test. To calculate the integral in (4) in the case of when $n = 10$ use the commands

> K10 := int(1/2^x, x = 10..infinity);

$$K10 := \frac{1}{1024\ ln(2)}$$

> evalf(");

$$.001408881876$$

Therefore $0.0014 - 2^{-10} \le R_{10} \le 0.00141$. In other words, the error in using s_{10} to approximate s is no larger than 0.00141 but no less than $0.0014 - 2^{-10}$.

Suppose we want to approximate the sum s and have an error of no more than 10^{-6}. If we use the partial sums as in (5), we need to take n large enough that $K_n \le 10^{-6}$. To determine when this happens, we use the commands

> Kn := int(1/2^x, x = n..infinity);
> n0 := fsolve(Kn = 1/1000000);

$$n0 := 20.46033494$$

This means that we need to take at least 21 terms in the partial sum to be able to guarantee the error (or remainder) is less than 10^{-6}. Now, suppose we use the approximation in (6). In this case we must take n large enough that $a_n/2 \le 10^{-6}$. Solving this by hand we find that $n \ge 6\ ln(10)/ln(2) - 1 \approx 18.9$. So the two methods end up requiring about the same number of terms. Note that the approximation in (6) can be calculated when $n = 19$, using the commands:

> approx := n -> sum(1/2^k , k = 1..n) + int(1/2^x , x = n..infinity) - (1/2^n)/2;
> evalf(approx(19));

As you will see in this Laboratory, the approximation in (6) can produce some very striking results in comparison with the approximation in (5).

Laboratory Problems

Problems 1 and 2 concern two applications of the ideas expressed in the preceding, and these problems are independent of one another.

1. In this problem you are asked to examine the series

$$s = \sum_{k=1}^{\infty} \frac{k}{1 + 2k^3}. \tag{7}$$

(a) Determine an appropriate function $f(x)$ for this series, and then calculate the integral in (4). Explain why this shows that the series converges.

(b) Plot the partial sums s_n for $1 \le n \le 50$ (to do this you may want to review the example in Lab 23). Why does this plot show that $R_1 > R_2 > R_3 > ... > R_{50} > 0$? Is it always true that $R_{n+1} < R_n$ for this series?

(c) How many terms of the series should be added together to guarantee that the remainder is no more than 10^{-4}? What value of n should be used to guarantee that the error in the approximation in (6) is no more than 10^{-4}?

(d) On the same graph plot the approximation for the sum given in (6) and the partial sums s_n for $1 \le n \le 50$. Which appears to converge to the value of the sum the fastest? Is your answer consistent with your values obtained in (c)?

2. In this problem you are to investigate the series

$$s = \sum_{k=1}^{\infty} \ln\left(\frac{e^{t/k}}{1 + t/k} \right). \tag{8}$$

This series may look a little odd but it is associated with an important function in mathematics known as the gamma function, denoted by $\Gamma(t)$. This function is related to s through the equation $\Gamma(t) = e^{s - \gamma t}/t$, where γ is Euler's constant ($\gamma = 0.5772156649...$). In Maple this number is designated as gamma (like π is designated as Pi). One interesting property of $\Gamma(t)$ is that $n! = \Gamma(n + 1)$ for $n = 1, 2, 3, ...$. Because of this, the gamma function is used to make sense out of expressions like $\pi!$ or $1.5!$ or $(-3.1)!$. In fact if you ask Maple, it will give you a value for these expressions and the way it is doing this is by evaluating the gamma function (can you guess what the value of $\pi!$ is?).

One thing that is different about the series in (8) is that the terms depend on the variable t. This means that the integral K_n in (4), which is used to estimate the error, also depends on t. The consequences of this are examined in the following.

(a) Suppose we want to calculate the sum of the series in (8) for values of t over the interval $0 \le t \le 2$. How many terms should be used to guarantee that the remainder R_n is less than 10^{-4} no matter what value of t is taken from this interval?

(b) Suppose we take $t = 1$. In this case $s = \gamma$. Let

$$\text{Error} := \text{gamma} - [\text{ approximation of } s \text{ given in (6) }]$$

and

$$\text{Error2} := \text{gamma} - s_n .$$

Calculate the value of these two errors for $n = 1, 10, 100, 1000, 10{,}000$. How many terms does it take before Error is less than 10^{-4}? Before Error2 is less than 10^{-4}? Based on your results, comment on the effectiveness of the approximations given in (5) and (6).

26

Calculating π

Objective
To investigate several ways of calculating approximate values of π , and to compare their levels of accuracy.

Background

One of the most important numbers in mathematics is the ratio of the circumference to the diameter of a circle. The symbol π that we use to designate this number was introduced early in the 18th century, and was popularized by Euler. The importance of π lies in its frequent appearance in mathematics, often in circumstances that are seemingly unrelated to circles.

   The numerical approximation of π is a problem that has concerned mathematicians from ancient times to the present. The foremost of the Greek mathematicians of antiquity, Archimedes, used inscribed and circumscribed polygons to estimate the value of π . With hexagons it is easy to show that $3 < \pi < 2\sqrt{3} \approx 3.46$. Archimedes used polygons with 96 sides to show that $223/71 < \pi < 22/7$; in decimal notation, this estimate becomes $3.1408... < \pi < 3.1428...$, from which we can conclude that $\pi = 3.14...$ to two decimal places.

In later times, many other mathematicians, including giants such as Newton and Euler, proposed methods yielding improved estimates. The discovery of calculus in the 17th century soon resulted in a host of formulas or relations from which approximate values of π can be calculated. One fellow, Ludolph van Ludolphian, calculated it to 35 decimal places, a feat that took him most of his life. He was proud enough of this accomplishment that he had the numbers engraved on his tombstone.

The development of high-speed computers in the late 20th century has made it possible to execute enormous computations. Consequently, billions of digits in the decimal representation of π are now known. Of course, these are far more digits than are needed for any practical computation, but the approximation of π remains today as a benchmark problem on which the efficiency of computers can be tested.

A related mathematical question is the determination of what kind of number π is, rational or irrational, algebraic or transcendental. It was demonstrated in the 18th century that π is irrational, that is, not a fraction. However, only in 1882 was it finally shown that π is transcendental, that is, not the root of any polynomial equation.

Two very readable references that deal with the numerical approximation of π are

(1) Petr Beckmann, **A History of** π **[Pi]**, Golem Press, New York, 1971.

(2) Jonathan M. Borwein and Peter B. Borwein, Ramanujan and Pi, **Scientific American**, Vol. 258, 112 –117, February 1988.

Laboratory Problems

The problems that follow describe several ways of approximating π. Some of these methods use formulas that occur in a calculus course, but others depend on more advanced mathematics. In each case you are asked to execute only a few steps of the process. Finally you are invited to draw some conclusions about the relative efficiency and accuracy of the various methods. It should be pointed out that the problems are independent and omitting any one will not affect your capability of doing any that follow. Also, in some parts of these problems you will need to set Digits to a number larger than the default value of 10.

1. The area of a circle of radius one is π. Expressing this fact in the form of an integral, we have

$$\pi = 4 \int_{0}^{1} \sqrt{1 - x^2} \, dx \; . \tag{1}$$

Therefore, approximating this integral by a Riemann sum yields an approximation to π.

(a) Use the Maple commands `leftsum` and `rightsum,` found in the `student` package, with $n = 100$, to approximate the integral in Eq. (1) (these commands are demonstrated in the example of Laboratory 9). Observe that `leftsum` gives an upper bound for the true value, while `rightsum` gives a lower bound. Find the average of the two sums. Note that this value gives an estimate that is in error by no more than one-half of the difference between the leftsum and the rightsum.

(b) Repeat the steps in part (a) for $n = 200$ and $n = 400$. How many decimal places in π are obtained correctly in each of these cases?

(c) Estimate the integral in Eq. (1) by using `trapezoid` and `simpson,` also found in the `student` package, with $n = 100$, 200, and 400. How many decimal places are found correctly in each case?

2. A famous relation involving π was discovered by Euler in 1736, namely,

$$\frac{\pi^2}{6} = 1 + \frac{1}{4} + \frac{1}{9} + \frac{1}{16} + \cdots + \frac{1}{n^2} + \cdots . \qquad (2)$$

Thus choosing a value for n, adding the squares of the reciprocals of the first n integers, multiplying by 6, and taking the square root, produces an approximation to π. Use Maple's `sum` command to handle the addition.

(a) Carry out the steps indicated for $n = 200$, 400, and 1000.

(b) Try to determine, or at least estimate, how many terms are needed to obtain a partial sum that approximates π with four decimal place accuracy. Note that the method of Laboratory 25 can also be used to estimate the sum of this series much more efficiently.

3. A third expression for π is Wallis' formula, discovered in 1655:

$$\frac{\pi}{2} = \frac{2}{1} \cdot \frac{2}{3} \cdot \frac{4}{3} \cdot \frac{4}{5} \cdot \frac{6}{5} \cdot \frac{6}{7} \cdot \frac{8}{7} \cdot \frac{8}{9} \cdots . \qquad (3)$$

To approximate π from this expression we need to carry out the indicated multiplication, terminating the process after a suitable number of terms. This multiplication is a little easier to execute if we group the terms in pairs. Then each pair of terms is of the form

$$\frac{(2k)(2k)}{(2k-1)(2k+1)} \quad \text{or} \quad \frac{4k^2}{4k^2-1} .$$

Maple's `product` command can be used to do the multiplication. Use Wallis' formula to approximate π, using 100, 200, and 400 terms.

4. In 1706 Machin discovered a formula for π that leads to much more efficient computations than those in the previous problems. Machin used this formula to compute π to 100 decimal places.

   (a) Machin's formula can be derived, using elementary trigonometry, as follows. Let $\theta = \arctan(1/5)$, so, $\tan(\theta) = 1/5$. Then, using the double angle formula for the tangent, show that $\tan(2\theta) = 5/12$ and $\tan(4\theta) = 120/119$. Now, since $\tan(\pi/4) = 1$ then it follows that 4θ is close to $\pi/4$. Using the difference formula for the tangent, show that $\tan(4\theta - \pi/4) = 1/239$. From this show that

   $$\frac{\pi}{4} = 4 \arctan(1/5) - \arctan(1/239) . \tag{4}$$

   To use Machin's formula (4) to approximate π it is also necessary to use the power series for the arctangent function, namely,

   $$\arctan x = x - \frac{1}{3}x^3 + \frac{1}{5}x^5 - \frac{1}{7}x^7 + ... = \sum_{k=0}^{\infty} \frac{(-1)^k x^{2k+1}}{2k+1} , \tag{5}$$

   which converges rapidly for values of x that are small compared with 1.

   (b) Use one term in the series in (5) for $\arctan(1/239)$ and four terms in the series in (5) for $\arctan(1/5)$, and then use Eq. (4) to calculate an approximate value for π. How many decimal places are given correctly in this way?

   (c) To obtain a better approximation use two terms in the series for $\arctan(1/239)$. How many terms should then be used in the series for $\arctan(1/5)$? To decide this you might want to try to make the errors in the computation of the two arctangents about the same. Then use Eq. (4) again to determine an improved estimate for π. How many decimal places are correct in your estimate?

   Note: There is an easy way to estimate the error in an alternating series such as in (5). Look in your calculus book if you don't remember how to do this.

5. An even more rapidly convergent series for the calculation of π was discovered by Ramanujan in 1914, namely,

   $$\frac{1}{\pi} = \frac{\sqrt{8}}{9801} \sum_{k=0}^{\infty} \frac{(4k)!(1103 + 26{,}390k)}{(k!)^4 396^{4k}} . \tag{6}$$

Unlike the formulas in the first four problems, this result depends on a knowledge of advanced mathematics, in particular, the area known as the theory of modular functions. Compute approximations to π by using one, two, three, and four terms, respectively, in the series in (6). How many decimal places are given accurately by each of these approximations?

6. Modular functions also underlie an iterative method recently discovered by J. Borwein and P. Borwein. The Borweins' method works as follows. To start out, let $y_0 = \sqrt{2} - 1$ and $\alpha_0 = 6 - 4\sqrt{2}$. Then define further values of y_n and α_n, for $n = 1, 2, 3, \ldots$, with the equations

$$y_n = \frac{1 - (1 - y_{n-1}^4)^{1/4}}{1 + (1 - y_{n-1}^4)^{1/4}} \; , \tag{7}$$

$$\alpha_n = [(1 + y_n)^4 \alpha_{n-1}] - 2^{2n+1} y_n (1 + y_n + y_n^2) \; . \tag{8}$$

What one does here, for successive values of n, is first to calculate y_n from (7) and then determine α_n from (8). (You may want to look over the examples in Laboratory 23 for a discussion of sequences that are defined iteratively.) One can prove that $\lim_{n \to \infty} \alpha_n = 1/\pi$, and therefore the values of α_n provide approximations to $1/\pi$. Compute α_0, α_1, and α_2, and then take their reciprocals to determine approximate values of π. How many decimal places are given correctly by each of these approximations?

7. Compare the methods you have used to approximate π, commenting on their accuracy and their complexity. Which do you think is the most interesting?

27

An Application of Taylor Polynomials: Approximation of an Arc Length Integral Containing a Parameter

Objective

To use Taylor polynomials to approximate an arc length integral and to explore the dependence of the arc length on a parameter.

The arc length L of a curve described by the polar equation

$$r = f(\theta) \ , \ \text{ for } \ \alpha \le \theta \le \beta \ , \tag{1}$$

is given by the integral

$$L = \int_{\alpha}^{\beta} \sqrt{[f(\theta)]^2 + [f'(\theta)]^2} \ d\theta \ . \tag{2}$$

In this Laboratory the function f also depends on a parameter a, and you are to investigate the nature of the dependence of L on a.

If the integral (2) cannot be evaluated in closed form, then one must resort to an approximation of some kind. One approach is to evaluate the integral numerically for various values of a. Another possibility is to replace the integrand by a simpler function, such as a Taylor polynomial, that can be more easily integrated. Both of these methods are explored here.

EXAMPLE

The Maple command `taylor` can be used to determine a Taylor series expansion. This command has three arguments and is of the form

> taylor(f, x = x0, n);

The first argument in this command is the function to be expanded, given by a Maple expression; the second is the center of the expansion; and the third determines the order of the remainder (or error). For example,

> T3 := taylor(exp(x), x = 0, 4);

$$T3 := 1 + x + 1/2\, x^2 + 1/6\, x^3 + O(x^4)$$

produces the Taylor series expansion of exp(x) about x = 0 through terms of degree three.

As seen in this example, Maple's response to the `taylor` command includes a remainder term of the form $O(x^n)$. It is necessary to eliminate this term, by means of the `convert` command, in order to execute further calculations or plots. For example, to remove it from T3 we would use the command

> P3 := convert (T3, polynom);

$$P3 := 1 + x + 1/2\, x^2 + 1/6\, x^3$$

This produces the third-degree Taylor polynomial of exp(x) expanded about x = 0. The `taylor` and `convert` commands can also be combined; for example,

> P3 := convert (taylor(exp(x), x = 0, 4), polynom);

Laboratory Problems

Consider the function r = f(θ) = 1 + a cos θ. You are to investigate how the arc length of the graph of f(θ) depends on the parameter a. In Problem 1 you are to set up the integral that determines L and then evaluate it for specified values of a. In Problems 2 and 3 you are to derive approximations for L as a function of a. Problem 3 depends on Problem 2, and Problem 2 depends on Problem 1.

1. (a) Draw the graph of this function for $0 \le \theta \le 2\pi$, using a = 0, a = 1, and at least one value of a between 0 and 1. Do not hand in these graphs. These curves are drawn to help you visualize the function you are to investigate in what follows.

(b) Set up an integral that gives the length L of this graph, that is, the perimeter of the region enclosed by the curve. The integral will depend on the parameter a as well as the integration variable θ.

(c) Evaluate L at a = 0 , 0.2 , 0.4 , 0.6 , 0.8 , and 1.0 . Plot the resulting values. Describe how you think L depends on a. Is L an increasing or decreasing function of a ? Is it concave up or concave down? To do this plot you may find some of the plots in the examples in Labs 23 and 24 helpful.

2. (a) Now we want to derive an approximate *formula* for L as a function of a by replacing the integrand in (2) with a Taylor polynomial in powers of a. Expand about a = 0 and keep terms at least up to a^6 (more if you wish). Integrate the resulting expression to obtain the desired formula for L. Denote this approximation as Lapprox and plot Lapprox , along with the values you obtained in Problem 1(c), for $0 \le a \le 1$.

(b) Test the accuracy of your approximate formula for L by evaluating it at a = 1 and comparing the result with the value calculated in Problem 1(c). Do you think the approximation is better or worse for smaller values of a ?

(c) Determine, at least approximately, the rate of change of L with respect to a when a = 0.5 . This may suggest one advantage of the approximate formula over the numerically calculated values.

3. (a) Another approximation for L results from expanding the integrand in Eq. (2) as a function of θ. Find the Taylor polynomial of degree six about θ = π/2 , and integrate this polynomial to obtain another approximation for L. Plot this expression for $0 \le a \le 1$ and compare it with the results from Problems 1(c) and 2(c).

(b) Which of the two approximate formulas—obtained in Problems 3(a) and 2(a)—is more accurate? Comment on the advantages and disadvantages of each method of approximation.

28

Exploring Taylor Polynomials

Objective

To investigate how a Taylor polynomial for a function changes with the degree of the polynomial and the point of expansion.

In this Lab, we will investigate how the Taylor polynomial

$$P_n(x) = f(x_0) + f'(x_0)(x - x_0) + \ldots + \frac{1}{n!} f^{(n)}(x_0)(x - x_0)^n \qquad (1)$$

of the function $f(x)$ expanded about the point x_0 changes with the degree n and the point of expansion x_0. Given a particular function, we will first keep x_0 fixed and then observe how $P_n(x)$ changes with n. Then, for a fixed value of n, we will try to determine the best point x_0 in the interval about which to do the expansion.

First, we give some examples to show how to use Maple to obtain Taylor polynomials, and we ask you some questions to start you thinking about how Taylor polynomials are used.

EXAMPLE 1

We begin by finding the second degree Taylor polynomial for $\sin(x)$ expanded about the point $x_0 = 5$. This function is entered into Maple with the command

> f := sin(x);

To obtain the Taylor series of f about $x_0 = 5$ having degree two enter the command:

> T2 := taylor(f, x = 5, 3);

$$T2 := sin(5) + cos(5)\,(x-5) - 1/2\,sin(5)\,(x-5)^2 + O((x-5)^3)$$

The last argument of this command specifies the order of the error. For the general formula for the Taylor series in (1), this error is $O((x - x_0)^{n+1})$. Since we want a polynomial with degree two then the order of the error is three, and so the last argument in the preceding `taylor` command is three.

Maple's response to the `taylor` command here includes the remainder term $O((x - 5)^3)$. To use other Maple commands with T2 it is necessary to eliminate this term, and to do so we use the `convert` command:

> P2 := convert(T2, polynom);

$$P2 := sin(5) + cos(5)\,(x-5) - 1/2\,sin(5)\,(x-5)^2$$

We could also have used the nested commands

> P2 := convert(taylor(f, x = 5, 3), polynom);

To see how the Taylor polynomial P2 compares with sin(x) over the interval [3 , 7] , use the plot command:

> plot({ f, P2 }, x = 3..7, title = `Plot for Example 1`);

At what point in the interval [3 , 7] do the two functions f and P2 differ the most? What is the absolute value of this greatest difference? At what point in the interval [3 , 7] do the two functions f and P2 differ the least? What is the absolute value of this least difference?

EXAMPLE 2

Next we find a higher degree Taylor polynomial for sin(x) expanded about the point $x_0 = 5$.

(a) Using commands similar to those in Example 1, find the Taylor polynomial P4 of degree four for sin(x) about the point $x_0 = 5$.

(b) Use the plot command

> plot({ f, P4 }, x = 3..7, title = `Plot for Example 2`);

to see how well P4 fits the function f(x) over the interval [3, 7] . At what point in the interval [3, 7] do the two functions f(x) and P4 differ the most? What is the absolute value of this greatest difference?

EXAMPLE 3

If a Taylor polynomial is going to be used to approximate a function over an interval, it is important to be able to measure the error in the approximation. One way to do this is to use the area bounded by the two curves. If the original function is f(x), the Taylor polynomial is P(x) , and the interval is $a \le x \le b$, then this error is given by the integral

$$\text{Err} = \int_a^b |f - P| \, dx .$$

This uses the difference between f(x) and P(x) . Another way to measure error is to use the square of this difference. This leads to the integral

$$\text{Error} = \int_a^b (f - P)^2 \, dx . \tag{2}$$

Let's see what values we get from these integrals in the case of when f = sin(x) and the interval is $3 \le x \le 7$. The point to be expanded about is $x_0 = 5$. In fact, let's calculate the errors for the Taylor polynomials of degree one, two, ..., five. These calculations can be automated using a do-loop as follows:

```
> for n from 1 to 5 do
>     P := convert( taylor( sin(x), x = 5, n), polynom );
>     Err[n] := evalf( int( abs(sin(x) – P), x = 3..7) );
>     Error[n] := evalf( int( (sin(x) – P)^2, x = 3..7) );
> od:
```

The next step is to put these values into lists that we can plot:

```
> LErr := [ [ j, Err[j] ] $ j = 1..5];
> LError := [ [ j, Error[j] ] $ j = 1..5];
```

Now, for the results:

```
> n := 'n':
> plot({LErr,LError},n=0.75..5,–0.3..2.2, title=`A Comparison of Errors`,
style=LINE);
```

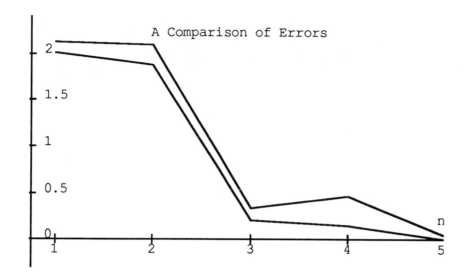

It is seen that both Err and Error show that the Taylor polynomial gets to be a much better approximation to sin(x) as the degree of the polynomial increases. The other observation to be made is that the calculations made by the do-loop take some time to complete. The reason is not the calculation of P or Error , but the calculation of Err . The fact that the absolute value is part of the definition of Err makes the calculation of the integral more difficult. This is because the integration formulas that Maple can use are limited if the absolute value function is present. It is for this reason that the error defined in Eq. (2) is preferable and the one we will use in the Lab problems.

Laboratory Problems

1. Consider the function

$$f(x) = \sin(x) + \exp(x) \quad \text{for } -6 \le x \le 2 \; .$$

We want to use a Taylor polynomial to obtain a "good" approximation of this function over the given interval. From the definition of the Taylor polynomial given in (1), it is clear that $P_n(x)$ and the function $f(x)$ are equal at the point of expansion x_0 . Thus it is reasonable to choose the midpoint of the interval $x_0 = -2$ as the point of expansion. However, it is not at all clear that the midpoint is the best point about which to do the expansion.

In this Laboratory problem, we investigate how to pick the expansion point of a Taylor polynomial of fixed degree so as to obtain the "best fit" to the function over the given interval. To determine this, the error defined in (2) will be used. For this problem the formula for this error takes the form

$$\text{Error} = \int_{-6}^{2} (f - P4)^2 \, dx \, . \tag{3}$$

(a) For the function $f(x) = \sin(x) + \exp(x)$, find the Taylor polynomial P4 of degree four expanded about the midpoint $x_0 = -2$ of the given interval. Plot $f(x)$ and P4 over the interval $[-6, 2]$.

(b) Calculate Error as given in (3). You should get an error of about 16.5 .

(c) Calculate Error in the case when P4 is determined using $x_0 = -2.5$. Now you should get an error of about 4.9 .

(d) For this example, the best point of expansion is not the midpoint, because the point $x_0 = -2.5$ gives a smaller error. To find the best point of expansion, we need to calculate symbolically a Taylor polynomial of degree four expanded about an arbitrary point $x_0 = a$. Use the nested commands

> P4a := convert(taylor(f , x = a, 5), polynom);

to obtain such a Taylor polynomial.

(e) Now Error will depend on the point of expansion $x = a$. Calculate this error using the command

> Error := int((f − P4a)**2, x = −6..2);

Do not be alarmed if Maple fills up several screens giving this error as a function of a .

(f) Find a value $a = a0$ so Error is minimized. Remember that a0 must be a point in the interval $[-6, 2]$. This a0 gives the best point in the interval about which to expand the fourth-degree Taylor polynomial.

(g) Construct a plot comparing f , P4 , and the fourth-degree Taylor polynomial using the best expansion point $x = a0$. Comment on how well the two polynomials approximate the function f . For example, we obtained P4a by minimizing the error in (3). How well does this approximation appear to do on a point-by-point basis in comparison with P4 ?

2. Consider a Taylor polynomial of degree five for the function $\sin(x)$ over the interval $[0, 2\pi]$. Let P5a denote the Taylor polynomial of degree five for $\sin(x)$ expanded about an arbitrary point a in the interval $[0, 2\pi]$.

(a) Find the point $a = a_1$ in the interval $[0, 2\pi]$ so that

$$\text{Error} \ = \ \int_0^{2\pi} (\sin(x) - \text{P5a})^2 \, dx$$

is minimized.

(b) In (a), you should have found that $a_1 \neq \pi$. Use the second derivative test to show that Error, as a function of a, has a local maximum at $a = \pi$.

29

Taylor Series Solutions of Differential Equations

Objective
To use Maple and Taylor polynomials to find approximate solutions to differential equations that determine the amount of money on deposit in a bank.

Taylor series are natural tools for approximating known functions. This Laboratory begins with a brief review of this process. However, Taylor series are useful in many other important mathematical applications. This Laboratory concentrates on one example of these, the problem of approximating solutions of differential equations. You will examine the familiar problem of money growth from continuous compounding of interest. You will see how to apply the Taylor series method in the simple case of an interest rate that is constant with time. Then you will apply the method to harder problems where the interest rate changes with time.

In this Laboratory you will use the `taylor` command to generate a series, along with the `convert` operation to convert a series to a polynomial. Illustrations are provided in the first example. The `collect` command is applied to group terms of like power in polynomials. Other commands that are used include the do-loop and the `sum` operation. The Maple `array` command is used to define coefficients of the series solution. For example,

```
> m := array(0..3);
> M := sum( m[i]*t^i, i = 0..3 );
```

would define an expression of the form

$$M := m[0] + m[1]t + m[2]t^2 + m[3]t^3 .$$

Finally the Maple differential equation solver `dsolve` with the "series" option is mentioned in case students wish to try this command, but no knowledge or application of it is necessary for the Laboratory.

This Laboratory is concerned with a practical use of Taylor series. You are already familiar with the idea of using Taylor series to approximate functions that are specified by formulas (see Labs 27 and 28). Here we extend that idea to functions that are specified as solutions of a differential equation (that is, no formula may be available for the function).

We briefly review the use of Taylor series to approximate a given function in Example 1.

EXAMPLE 1

Consider the function $f(t) = e^{6t/100}$ on the interval $[-10, 10]$:

```
> f := exp(6*t /100);
```

In these examples we suppress Maple's responses; however, you should try the commands to see what the responses are. The Taylor series for f about t = 0 containing terms up to powers of t^2 is constructed using the command

```
> T2 := taylor( f, t = 0, 3);
```

Maple includes the order of the remainder term, which in this case is $O(t^3)$. To obtain the corresponding Taylor polynomial, P2 , we remove the remainder term using the `convert` command:

```
> P2 := convert( T2, polynom );
```

We urge you to perform both of these steps together:

```
> P2 := convert( taylor( f, t = 0, 3), polynom );
```

We can construct, and then plot, several different Taylor approximations to f as follows:

```
> for n from 3 to 7 do
>     P := convert( taylor(f, t = 0, n), polynom);
>     plot({f, P}, t = -10..10, title = `f(t) and Taylor Polys of f(t)`);
> od;
```

The error in the approximations can also be examined using the do-loop:

```
> for j from 3 to 7 do
```

```
>    Error := f − convert( taylor(f, t = 0, j), polynom);
>    plot(Error, t = −10..10, title = `Error in Taylor Polys of f(t)`);
> od;
```

Notice that the maximum error decreases (roughly) by a factor of 10 with each additional term in the Taylor series. This raises several questions that you should think about, including:

    (i)   What degree of Taylor polynomial is necessary to approximate f(t) within 1% throughout the interval?

    (ii)   Where is the largest error, and why does it occur there?

Next we construct Taylor series approximations of functions that are solutions of differential equations.

EXAMPLE 2

Suppose an amount of money M(t) at time t increases at a 6% rate of interest, compounded continuously. Suppose also we start (at t = 0) with $1000, or 1 K$ (in units of kilodollars). The equation that M(t) satisfies is

$$M'(t) - \frac{6}{100} M = 0 \ . \tag{1}$$

This is entered into Maple with the command

```
> eq := Mp − 6*M/100 = 0;
```

where Mp stands for the derivative M'(t) .

    We also know the solution of this problem; it is just the function f(t) from Example 1. For the moment, let us pretend that we **do not know** this, and let us derive a Taylor series approximation to it. Why would we ever wish to do this? Clearly, it is unnecessary for this simple problem. The purpose is to illustrate a method that also works for more difficult examples, for which the solution is not so obvious.

    To begin, we first define a Maple array that contains a collection of subscripted variables that will be the coefficients in the Taylor series of the solution M :

```
> m := array(0..6);
```

We can now refer to quantities such as m[0] and m[3] (but not m[7]). The Taylor series in powers of t , through t^6 , can be written

```
> Mapprox := sum( m[i]*t^i, i = 0..6);
```

Notice next that the initial condition M(0) = 1 applies to Mapprox . This condition comes from the fact that we begin, at t = 0 , with $1000 (or 1 K$). So, Mapprox = 1 at t = 0 . This means that m[0] = 1 , and we enter this into Maple:

```
> m[0] := 1;
```

The derivative of the approximation is just

> Mapproxprime := diff(Mapprox, t);

We now insert Mapprox and Mapproxp into Eq. (1) for M that we called eq:

> eq1 := subs(M = Mapprox, Mp = Mapproxp, eq);

There are a bunch of terms in eq1 in powers of t , and having coefficients involving m[1] , ..., m[6] . We next group all the like powers of t , using the command

> eq2 := collect(eq1, t);

     Now comes an important mathematical step. Notice that eq2 is an equation of sixth degree (in powers of t) that is equal to zero. We want this equation to hold for **all values** of t . Therefore the coefficient of **each power** of t must vanish. (Convince yourself of this statement before going further.)
     How can we use Maple to set each of the coefficients to zero? If we first set t = 0 in eq2 , then what remains are the constant terms equal to zero. If we differentiate eq2 once and set t = 0 , then what remains is the coefficient of t equal to zero. If we differentiate eq2 twice and set t = 0 , then what remains is the coefficient of t^2 equal to zero...and so on. If we continue this up through the fifth derivative of eq2 , we obtain a total of six equations. We can solve these six equations for the six quantities m[1] , m[2] , ..., m[6] . Here is a set of Maple commands that does these steps:

> m[1] := solve(subs(t = 0, eq2), m[1]);
> for i from 2 to 6 do
> m[i] := solve(subs(t = 0, diff(eq2, t $ (i–1))), m[i]);
> od;

We now have obtained all the coefficients of the approximation to our solution.

> Mapprox;

Of course, we **know** that the solution to this simple problem is the function f of Example 1. If the method that we have been using is indeed correct, we would think that our answer Mapprox should be a Taylor series approximation to f . Is it?

> Mapprox – convert(taylor(f, t = 0, 7), polynom);

                                  0

Some questions:

    (i)   How much money is available after 10 years, according to the approximate solution? According to the exact solution?

(ii)   Think back to the equation eq2 , which we said was sixth degree in t .
What would happen if we defined one more "coefficient equation," by
computing the sixth derivative of eq2 ? Convince yourself that you
would find the equation m[6] = 0 , which contradicts the result for m[6]
that was found previously. What causes the problem here? When you
use the series method to find approximate solutions, you must be alert to
this kind of problem. You must always be sure that you have included
all terms of a given power of t before you differentiate and solve the
coefficient equation.

(iii)   How would we proceed if we wanted a more accurate approximate
solution, for example, one containing powers of t through the eighth
degree?

Laboratory Problems

In the Lab problems you are to use Taylor polynomials to find accurate
approximations to the solutions of a "money equation," such as the one considered
in Example 2. In Problem 1 a time varying interest rate is considered, and in
Problem 2 you are to investigate the situation when money is also being deposited
to the account as time passes. Although it is not mandatory, it is recommended that
Problem 1 be done before Problem 2.

1. Suppose the rate of continuous compounding of interest is not constant
over time. Specifically, suppose the interest rate undergoes a decline over
the 10-year period, according to the formula

> k := 6/100 + (2/100)*cos(Pi*t/20);

Then the problem to be solved is the same as Eq. (1) in Example 2 but with
the coefficient 6/100 replaced by the function k that now depends on t .
The solution to this new differential equation is not in the form
M = exp((constant)*t) . Therefore we use series methods to find an
approximation to the solution.

(a) First find a Taylor series approximation, kapprox , to the interest rate
k , including terms through t^6 . How accurate (be specific) is this
approximation for t in [0 , 10] ? Since kapprox is a very good
series approximation to the coefficient in the differential equation, this
suggests (**but does not prove**) that we could find a very good series
approximation to the solution.

(b) Now we want to **follow the procedure in Example 2** as much as possible. Define a series approximation for M and its derivative, and insert both of these formulas, along with kapprox , into the equation.

(c) Collect all like powers of t . Solve for the coefficients in the Taylor series approximation of the solution.

(d) Plot your approximation to the amount of money as a function of time, t , over the 10-year interval. What is the amount after 10 years? Another procedure for determining M(10) , which you need not know about now, produces the result $2069.53 . How close is your answer to this value?

(e) There is another comparison you can make. It is possible to show that the amount of money after exactly 10 years is the same as if the interest rate were the constant value kav = 0.07273239544 . That is, the function Mav = exp(kav*t) has the value 2.06953... after 10 years. Plot, over the 10-year interval, k and kav together on the same graph. On another graph, plot Mav and your approximation to M . Comment on any differences.

2. Suppose that, in addition to the variable interest rate in Problem 1, you (continuously) add to the amount of money M(t) at a constant rate of A kilodollars per year. The amount at any time t satisfies the equation

$$M'(t) - k(t)M = A .\qquad(2)$$

(a) Once again, use the series method to obtain an approximation for M(t). You can do this in either one of two ways.

Option 1: Proceed just as in 1 (a)–(c), including the extra A term from the start and using Eq. (2).

Option 2: Use the Maple command `dsolve` and the "series" option. You can learn how to use this command from the help screen. **Warning:** The syntax for `dsolve` must be followed carefully. You may find that **Option 1** requires learning fewer new Maple commands, but requires more entering of commands, than **Option 2.**

(b) After you have obtained a series approximation for M , answer the following question: How big should the annual payment rate A be so that you have $8000 after 10 years ?

30

Graphs in Polar Coordinates

Objective

To plot some interesting curves described by simple functions in polar coordinates, and to see how these curves depend on one or more parameters in the defining function.

EXAMPLE

Plot the graph in polar coordinates of the equation

$$r = 1 - a \sin(5t)$$

for $t = 0..2*Pi$, and for values of a between 0 and 2 . Note that we are using t for the polar angle θ .

   To produce a plot in polar coordinates, we first define the function:

```
> f := 1 − a*sin(5*t);
```

Next we assign a value to the parameter a and issue the following plot command:

```
> a := 0.5; plot( [ f, t, t = 0..2*Pi], coords = polar);
```

Note the square brackets, which request a parametric plot, and the option 'coords = polar', which specifies that the plot is to be drawn in polar coordinates.

To get a sense of how the graph of this function f changes as the parameter a changes, one can write a do-loop to plot several graphs in fairly rapid succession. For example:

```
> a := 'a';
> for n from 0 to 4 do
>     plot( [subs(a = 0.5*n, f), t, t = 0..2*Pi], coords = polar);
> od;
```

The graphs for a = 0 , 2 , and 4 are shown here.

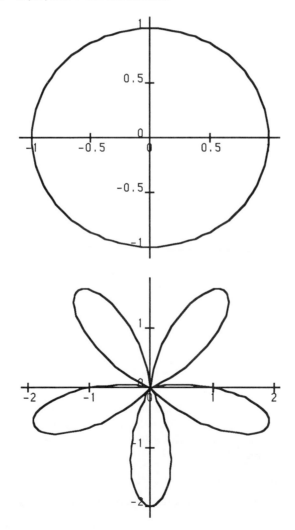

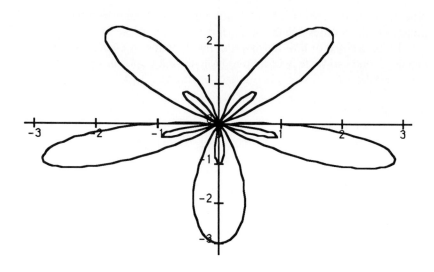

Laboratory Problems

Each of the following problems calls for you to plot graphs of a function in polar coordinates. The functions depend on one or more parameters, and you are to determine how the graph changes as the parameters are varied. Your instructor will tell you how many plots, if any, are to be handed in.

1. Let $r = 1 + a \cos(2\theta)$. Plot the graph of this function for $0 \le \theta \le \pi$ and for several values of a between 0 and 2. Describe in your own words how the graph changes as a increases. Pay particular attention to the neighborhood of the point $a = 1$.

2. Let $r = a \cos(\theta) + \sin(b\theta)$.

   (a) Plot the graph of this function for $0 \le \theta \le \pi$ with $a = 2$ and $b = 12$.

   (b) With $b = 12$, what is the effect on the graph of using values of a larger than 2? What is the effect of using values of a that are smaller than 2 but nonnegative? What is the effect if a is negative?

   (c) With $a = 2$, plot the graphs corresponding to $b = 11$, $b = 23/2$, and $b = 23/4$. Observe that in the last two cases the graph is not a closed curve on the interval $0 \le \theta \le \pi$. What can you do to cause the graph to form a closed curve?

3. Let $r = 1 + a \cos(b\theta)$.

   (a) Plot the graph of this function for $0 \le \theta \le 2\pi$ with $a = 0.2$ and $b = 6$.

(b) With $b = 6$, let a take on increasing positive values and plot the graph in each case. Describe how the graph changes as a increases. In particular, what happens as a passes through one?

(c) Let $a = 0.6$ and $b = 13/2$. Plot the graph in this case, choosing a θ interval to make the graph a closed curve. Then let $a = 3$, and redraw the graph in this case. Describe the differences.

4. Let $r = 1 - a \sin \theta + b \sin 3\theta$.

(a) Let $a = 0$ and let b increase from 0 to 2. Plot the graph for several such values of b, and describe how the graph changes as b increases.

(b) Repeat part (a) for $a = 1/2$, $a = 1$, and $a = 2$. How does the evolution of the graph as b increases differ in these three cases?

31

Introduction to Surface Plotting

Objective
To learn how to use Maple to plot surfaces in three dimensions, including surfaces described by Cartesian or by parametric equations.

The purpose of this Laboratory is to give you some practice in using Maple to plot surfaces in three dimensions. We will do this by using what are known as quadric surfaces; they are the counterparts in three dimensions of the conic sections in two dimensions. You may want to refer to your textbook to familiarize yourself with the mathematical formulas for these surfaces. In this Lab you will also be introduced to a parametrization of a surface. This is needed as it enables us to draw much smoother surfaces.

EXAMPLE

Consider the equation

$$2x^2 + y^2 + 3z^2 = 1 .$$

This represents a surface in three-dimensional space, and in this example we will show how to use Maple to plot such a surface.

First we show what **not** to do. Enter the equation as follows:

> eq := 2*x^2 + y^2 + 3*z^2 = 1;

Maple's plot command in three-dimensions is `plot3d,` and we try to use it to plot the surface over $-2 \le x \le 2$ and $-2 \le y \le 2$:

> plot3d(eq,x=-2..2,y=-2..2);

Error, (in plot3d)
1st argument must be an expression, a procedure or a list of 3 of these

We get an error message, and the reason is that Maple plots functions and not equations. One way to proceed is to solve the given equation for z as a function of x and y:

> soln := solve(eq, z);

$$soln := 1/3\ 3^{1/2}\ (-2\ x^2 - y^2 + 1)^{1/2}\ ,\ -1/3\ 3^{1/2}\ (-2\ x^2 - y^2 + 1)^{1/2}$$

This gives us two solutions. Let's call them z1 and z2:

> z1 := soln[1]; z2 := soln[2];

$$z1 := 1/3\ 3^{1/2}\ (-2\ x^2 - y^2 + 1)^{1/2}$$
$$z2 := -1/3\ 3^{1/2}\ (-2\ x^2 - y^2 + 1)^{1/2}$$

We can now plot z1 and z2:

> plot3d(z1, x = -1..1, y = -1..1);
> plot3d(z2, x = -1..1, y = -1..1);

Or we can plot them together. This is accomplished as follows:

> plot3d({z1, z2}, x = -1..1, y = -1..1);

This is better but not great since some of the graph appears to be missing. The reason for this lies in the rectangular grid system Maple uses to plot the surface. One way to cope with this problem, and to obtain a better plot of the surface, is to use a parametric description of the surface.

Since the surface is two dimensional, we are going to need two parametric coordinates. We will denote these coordinates by u , v . For the given surface, a parametric representation is the following:

$$x = \frac{1}{\sqrt{2}}\ \cos(u)\ \cos(v)\ ,\ y = \cos(u)\ \sin(v)\ ,\ z = \frac{1}{\sqrt{3}}\ \sin(u)\ .$$

In this case $0 \le u \le 2\pi$ and $0 \le v \le \pi$. It is not difficult to verify that these expressions satisfy the original equation, and this will be discussed further in what follows.

In the following we will use f, g, and h rather than x, y, z.

```
> f := cos(u)*cos(v)/sqrt(2);
> g := cos(u)*sin(v);
> h := sin(u)/sqrt(3);
```

Now we'll plot the surface:

```
> plot3d( [f, g, h], u = 0 .. 2*Pi, v = 0 .. Pi, title = `2x^2 + y^2 – 3z^2 = 1`);
```

After the surface is plotted rotate it so Theta = 50 and Phi = 85. This egg shaped surface is known as an *ellipsoid*. What is significant is that by using the parametric description of the surface we have produced a much smoother graph.

As stated earlier, it is not difficult to check that $2f^2 + g^2 + 3h^2 = 1$ (i.e., the parametrization satisfies the original equation). This means that x = f(u ,v) , y = g(u ,v) , z = h(u ,v) is a solution of the equation. The harder question to answer is what range of values should u and v have. Try experimenting, using `plot3d,` with different intervals for these parameters. For example, you might try:

(i) $0 \le u \le \pi$, $0 \le v \le \pi$,

(ii) $\pi/2 \le u \le 3\pi/2$, $0 \le v \le 2\pi/3$, and

(iii) $0 \le u \le 3\pi$, $0 \le v \le \pi$.

From this you should conclude that if smaller ranges are used then the graph is incomplete, and if larger ranges are used then nothing new appears.

Laboratory Problems

1. Consider the equation $z^2 = 1 + x^2 + y^2$. Solve the equation for z (you can do this by hand if you wish), and then plot both positive and negative square roots on the same graph for x = –2..2 , y = –2..2. Change your viewing point so that Theta = 70 (degrees) and Phi = 80 (degrees).

2. Consider the equation $x^2 + y^2 – z^2 = 1$.

   (a) Solve the equation for z (you can do this by hand if you wish), and then plot both positive and negative square roots on the same graph for x = –2..2 , y = –2..2. Observe that the graph is not well drawn near the plane z = 0.

(b) A parametrization for this surface is

$x = \cosh u \ \cos v, \ \ y = \cosh u \ \sin v, \ \ z = \sinh u,$

where $-1 \leq u \leq 1$ and $-\pi \leq v \leq \pi$. It is not hard to show that these expressions satisfy the given equation if you recall that $\cosh^2 u - \sinh^2 u = 1$. Plot the graph of these equations for $u = -1..1$, $v = -Pi..Pi$. Change your viewing point so that Theta = 30 and Phi = 60.

3. Consider $z = x^2 - y^2$. Plot the graph of this function for $x = -2..2$, $y = -2..2$. Parametric equations are not needed in this case. Drag the graph by moving the cursor so as to see the graph from several viewing points. Display the surface in an orientation that you think makes the shape of the graph clearly understandable. Also include the axes and identify which axis is which.

4. Consider the equation $x^2 - y^2 - z^2 = 1$. Construct a set of parametric equations for this surface (they should be rather similar to those in Problem 2). Then plot the surface in an orientation that makes its shape clear.

32

Surface Plotting and Contour Maps

Objective
To use Maple to plot surfaces in three dimensions and to understand how they relate to their contour maps.

The purpose of this Lab is to give you more practice in using Maple to plot surfaces in three dimensions. It is also intended to help you make sense out of contour maps, that is, the level curves associated with a surface.

In preparation for this Lab you may want to review the example of the previous Laboratory, which discusses the use of the `plot3d` command.

EXAMPLE 1

To illustrate the ideas developed in this Laboratory consider the function

$$z = y(1 + x^2 + y^2)e^{-x^2 - y^2} .$$

The surface plot and contour map for this function are shown in the following figures (for $-3 \le x \le 3$ and $-3 \le y \le 3$). There is a local maximum and a local minimum, and these are clearly seen in the surface plot. They are also evident in the contour map since they are the centers of the closed curves. However, the contour map does not indicate which is a min and which is a max since the level curves are not labeled.

The level curves in this contour map, as well as in the others in this Lab, are constructed using a fixed increment along the z-axis (the increment changes, however, if we change the function). Suppose for this example the increment is Δz. This means that if you pick a curve from the contour map, and it corresponds to the points on the surface that are at height $z = c$, then any curve adjacent to it on the contour map corresponds to either $z = c + \Delta z$ or $z = c - \Delta z$.

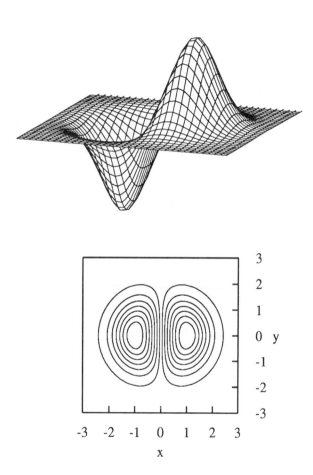

EXAMPLE 2

A different situation arises with the function $z = x^2 - y^2$. This does not have a local max or min, but it does have a saddle point, which occurs at $x = y = 0$. The contour map, for $-3 \le x \le 3$ and $-3 \le y \le 3$, is shown in the following figure. In this case, the curves are not closed like those that encircle the local max and min points in the last example. Near the saddle point they have the appearance of hyperbolas. Each set of hyperbolas corresponds to either a valley or hill (the reason for this structure near a critical point will be explored in Lab 36). You should use `plot3d` to plot the surface and convince yourself that the hyperbolas in the contour map that open to the left, as well as those that open to the right, correspond to hills. Similarly, those that open upwards, and those that open downwards, correspond to valleys.

For those who may be interested, the contour maps in this Laboratory were drawn with the program `gnuplot,` which is available at no charge via Internet using anonymous ftp from dartmouth.edu .

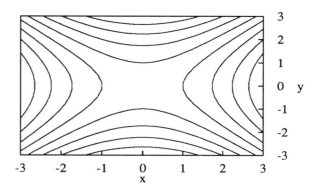

Laboratory Problems

In this Laboratory you will be working with the following surfaces:

(i) $z = \left(1 + \cos(10\sqrt{x^2 + y^2})\right)e^{-x^2 - y^2}$,

(ii) $z = (5x^2 + 2y^2)e^{-x^2 - y^2}$,

(iii) $z = y(1 - 10xy)e^{-x^2-y^2}$,

(iv) $z = y^3 - 3yx^2$,

(v) $z = \cos(x/2)\sin(y)$,

(vi) $z = 2x^4 - x^2 - 3y^2$.

On the next two pages you will find plots of these surfaces along with their contour maps. You are to use Maple to determine which surface corresponds to which plot and contour map.

1. Plot each of the surfaces so it resembles one of the plots on the next page. It is not necessary that your plots look exactly like the plots on the next page but there should be a clear resemblance between them.

2. In conjunction with each surface plot, identify which contour map is associated with it. Give a **clear and coherent** justification for your choice. This includes labeling, and classifying, the critical points (in the case of saddle points you should identify the hills and valleys in the contour map). You should also draw in the x-axis and y-axis on the contour maps.

a)

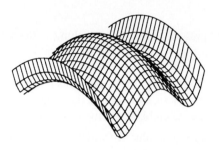

b)

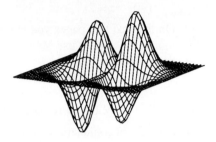

c)

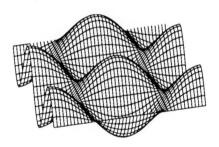

d)

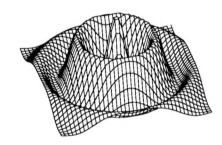

e)

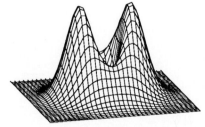

f)

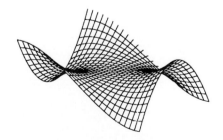

A)

B)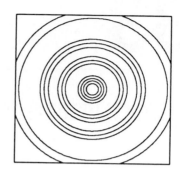

C)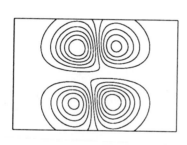

D)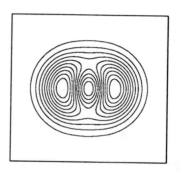

E)

F)

33

Introduction to Vector Analysis

Objective

To learn how to use Maple to manipulate vectors and vector functions. These are used to construct vector parametric equations for lines.

Many of the Labs in this book require the introduction of only one or two new Maple commands in order to carry out the necessary calculations. This Laboratory is somewhat different as it introduces you to using vectors in Maple, and this will require several new commands. It is strongly recommended that you spend time with the examples to familiarize yourself with these commands before proceeding with the Laboratory problems.

EXAMPLES

To define the vectors $\mathbf{r}_0 = (1, 0, 1)$ and $\mathbf{s}_0 = (-1, -1, 2)$ in Maple we first read the commands in the linear algebra package and then use the vector command:

```
> with(linalg):
> r 0 := vector( [ 1, 0, 3 ] );
> s 0 := vector( [ -4, -1, 2 ] );
```

The linear algebra package contains the commands `vector,` `norm,` `angle,` and `dotprod` (these are described later), along with many others. Consequently, this package must be read **once** every session to be able to use these commands.

It is easy to manipulate vectors in Maple. For example, to find $z_0 = 2r_0 - \frac{1}{5}s_0$ use the command

> z0 := evalm(2*r 0 – s0/5);

To find the dot product $m = r_0 \cdot s_0$ enter

> m := dotprod(r 0, s 0);

and to find the angle between these vectors use the command

> angle(r 0, s 0);

The length (or norm) of the vector r_0 is obtained using the following:

> d := norm(r 0, 2);

The 2 in this command indicates to Maple which norm it should use. There are many ways to measure a length of a vector, but we will stick to the standard Euclidean length (which is what the 2 indicates). If you are curious, you might try replacing the 2 with a 1 or infinity and see if you can determine how the norm is calculated. Also, as a note of caution, do **not** use the word "length" as a variable, because Maple has reserved this for its own use.

In this Lab we are interested in curves in three dimensions. As an example, to enter $r = (\cos(t), \sin(t), t)$ use the vector command:

> r := vector([cos(t), sin(t), t]);

It shouldn't be surprising that any of the previous vector commands can be used on such a vector. What is a little more involved, however, is evaluating r at particular values of t. To illustrate, suppose we want to evaluate r at $t = 1$. This is accomplished using the following command:

> subs(t = 1, op(r));

What is new here is the $op(\cdot)$ function.

Another slight complication with vectors arises if you want to calculate the tangent vector r'. This is accomplished using

> dr := map(diff, r, t);

The unexpected part of this command is the map(·) operation.

We will also need to be able to plot curves, and this can be done using the `spacecurve` command, which is contained in the `plots` package. To demonstrate, suppose we want to plot the curve **r** previously given for $0 \le t \le 4\pi$. The commands that accomplish this are

```
> R := convert( r, list);
> with(plots):
> spacecurve( R, t = 0..4*Pi );
```

The first command is needed to convert the vector **r** into a list that `spacecurve` will understand. Unfortunately, in Maple V, Release 1, the `spacecurve` command doesn't work on some machines (or it doesn't work with the `display3d` command introduced later). However, there is another way to plot curves in three dimensions using `plot3d.` For example, to plot **r** you can use the commands

```
> R := convert( r, list);
> plot3d( R, t = 0..4*Pi, s = 0..1, grid = [35, 2]);
```

In the `plot3d` command, the variable s is a "dummy" and has no effect on the plot (however, it is necessary to have such a dummy variable for this command to work). The "grid" option specifies how many points are to be used over the t interval (35) and over the s interval (2). The number for s is low because we don't want Maple to do unnecessary work for this dummy variable.

It is sometimes of interest to be able to plot two curves on the same graph. For example, suppose we want to plot **r** along with the straight line $\mathbf{s} = (1 + t, 1 - t, t)$ for $-1 \le t \le 3$. To put these on the same graph enter the following:

```
> F := plot3d( R, t = 0..4*Pi, s = 0..1, grid = [70,2] ):
> G := plot3d( [ 1 + t, 1 - t, t ], t = -1..3, s = 0..1 , grid = [10,2] ):
> with(plots):
> display3d( { F, G } , title = `A Couple of Curves` );
```

Note the colon (:) following the `plot3d` commands for F and G. This is strongly recommended to prevent a lot of unnecessary information from appearing on the screen. Also the command `display3d` is in the `plots` package, and this must be read in before using `display3d.` As a final comment, the difference in the grids for F and G is intentional, since one is a straight line while the other isn't.

Laboratory Problem

1. The distance between a curve $r(t)$ and a point x_0 is defined as the smallest value of $f(t) = \|r(t) - x_0\|$. In other words, the distance between a curve and a point x_0 is the shortest distance between x_0 and the points on the curve. This quantity arises frequently in applications. For example, it plays a central role in regression analysis of experimental data. In this Lab you are to use Maple to calculate the distance between a point and a curve.

    In the following we will take

    $$r = (\, t\,\cos(10t)\,,\, t\,\sin(10t)\,,\, t\,)\ ,\ \text{for } 0 \le t \le \pi,$$

    and

    $$x_0 = (\,1\,,\,1\,,\,1\,)\ .$$

    (a) Plot $f(t)$ and from this determine an interval $t_0 < t < t_1$ that contains the minimum point (denote this point as t_c). This interval should contain no other local maximum or minimum points for $f(t)$.

    (b) Find t_c and then determine the distance between the curve r and the point x_0 . Also, find the point r_c on the curve that determines this point (i.e., find $r(t_c)$). In your write-up explain why it is necessary to find the interval in (a) to be able to find t_c .

    (c) Find a vector parametric equation for the line that passes through the points x_0 and r_c .

    (d) Find a tangent vector to the curve at r_c and a vector parametric equation for the tangent line.

    (e) Plot, on the same graph, the curve r (remember $0 \le t \le \pi$), the straight line you found in (c), and the tangent line you found in (d). In each `plot3d` command set view = [−Pi..Pi, −Pi..Pi, 0..Pi] .

    (f) Determine the angle between the tangent vector and the vector $a = x_0 - r_c$. Express your answer in degrees. Do you think this value is a fluke, or do you think this will always be the case? For example, if you were to take another x_0 and redo this problem, do you think the angle would turn out to be the same? Give a mathematical, or geometrical, explanation why you think this happens.

34

Kepler's Laws and Planetary Motion

Objective
To use Maple to examine the motion of the planets and comets around the sun.

In this Laboratory you are asked to investigate some of the consequences of Kepler's laws. Before stating the Lab problems, the laws and some other relevant information are discussed. The presentation is brief, and you should look over the section in your textbook on this material for a more thorough treatment.

Kepler's first law states that planetary orbits around the sun are elliptical and can be described using the polar equation

$$r = \frac{a(1 - e^2)}{1 + e\cos\theta} \, .
\tag{1}$$

This equation places the sun at the origin of the coordinate system, and the origin is also one focus of the ellipse (see figure following). In this equation, a is the semimajor axis of the ellipse and

$$e = \sqrt{1 - (b/a)^2}$$

is the eccentricity, where b is the semiminor axis of the ellipse. One last comment: Although (1) was originally stated for planets, we will also be using it to describe the motion of comets.

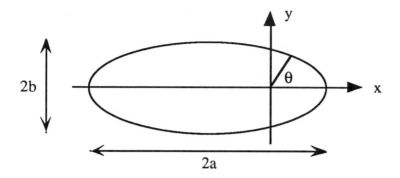

Kepler's second law concerns the rate at which area is swept out by the planet (or comet). Of specific interest to us is the velocity of the object, and for this we need the following formulas:

$$\frac{d\theta}{dt} = \frac{h}{r^2} \quad \text{and} \quad \frac{dr}{dt} = \frac{aeh}{b^2} \sin\theta , \tag{2}$$

where $h = 2\pi ab/T$, and T is the period of the orbit.

To add a little ambience to this problem, we will introduce some terminology used in celestial mechanics. The point of the orbit that is closest to the sun is called the perihelion, and the point farthest away from the sun is the aphelion. The angle θ, which is just the usual polar coordinate angle, is called the true anomaly. Also, it is interesting that Halley's comet has a retrograde orbit. This means that it rotates clockwise while the plants rotate counterclockwise. Finally, we will be measuring distances in astronomical units (AUs), where 1 AU = 149.6 x 10^6 km.

The last thing we need for the Lab is some data and that is given in the table that follows. The parameter ω that appears in the table is used in Problems 2 and 3 of the Lab. Also, the values for the semiminor axis are not given in the table, since they can be calculated using a and e. One last item: The mass of Halley's comet is approximately 2 x 10^{14} kg, while Earth's is 6 x 10^{24} kg.

	a (in AU)	e	ω (in degrees)	T (in Earth years)
Earth	1	0.02	103	1
Jupiter	5.2	0.05	14	12
Neptune	30.1	0.01	44	165
Pluto	39.4	0.25	223	248
Halley's comet	17.9	0.97	112	76

Laboratory Problems

The three problems are independent of one another. In Problems 1 and 2 the solar system is considered two dimensional, and in Problem 3 it is three dimensional. Problem 3 is somewhat more challenging than the others.

1. (a) On the same graph, plot the orbits of the objects listed in the preceding table. Identify the planets and comet on the graph (you will have to do this by hand).

   (b) Using the x,y-coordinate system, the velocity vector is

   $$\mathbf{v} = (\frac{dx}{dt}, \frac{dy}{dt}) \ .$$

   The speed S of the object is the length of this vector. The complication here is that Kepler's laws give everything in terms of polar coordinates. Using the fact that $x = r \cos \theta$ and $y = r \sin \theta$, and using the chain rule, show (either by hand or using Maple) that

   $$S^2 = (\frac{dr}{dt})^2 + r^2(\frac{d\theta}{dt})^2 \ .$$

   With this and the formulas in (1) and (2) express S in terms of θ. From this determine at what value(s) of θ the speed S has a maximum and at what value(s) it has a minimum. It should be remembered here that $0 \le e < 1$ for this problem.

   (c) The potentially catastrophic effects of a comet hitting Earth has been the subject of fables and theories since antiquity. For those who may be interested in some of the more wild ideas that have been connected with comets, the book **The Origin of Comets** by M. E. Bailey, S. V. M. Clube, and W. M. Napier, is recommended. In any case, as seen in the plot in (a), Halley's comet crosses Earth's orbit twice. Determine the values of θ when this happens. How fast is each object going when they plow into each other?

2. Even though the paths of the planets are elliptical they do not share a common axis as was assumed in Problem 1. If this is taken into account then in the x,y-plane the position vector $\mathbf{R} = (x1, y1)$ for the planet (or comet) has components

$$x1 = r \cos(\theta + \omega) \ , \tag{3a}$$

$$y1 = r \sin(\theta + \omega) \ , \tag{3b}$$

where r is still given in (1) and ω is a constant. The value of this constant for each object is given in the preceding table (also, remember that trigonometric functions in Maple require angles measured in radians).

(a) On the same graph, using (3a ,b), plot the orbits of the objects listed in the previous table. Identify the planets and comet on the graph (you will have to do this by hand).

(b) Halley's comet still crosses Earth's orbit twice. Determine the values of θ when this happens.

3. We now consider the orbits in three dimensions. To explain this situation, recall that a planet's orbit lies in a plane, and within that plane the orbit is elliptical (this is Kepler's first law). The difficulty is that the planes for the planets are different.

We need to pick a coordinate system, and it is usual in celestial mechanics to have the sun at the origin and have the x,y-plane be the plane that contains Earth's orbit (what astronomers call the ecliptic plane). In this case, given any particular planet or comet, its position vector $\mathbf{r}$ = (x2 , y2 , z2) has components given as

$$x2 = r\left(\cos(\Omega) \cos(\theta + \omega) - \sin(\Omega) \sin(\theta + \omega) \cos(i) \right) \qquad \text{(4a)}$$

$$y2 = r\left(\sin(\Omega) \cos(\theta + \omega) + \cos(\Omega) \sin(\theta + \omega) \cos(i) \right) \qquad \text{(4b)}$$

$$z2 = r \sin(\Omega) \sin(\theta + \omega) \sin(i) , \qquad \text{(4c)}$$

where r is given in (1), ω is given in the previous table, and Ω and i are constants that depend on which planet or comet we consider. Their values are given in the table that follows. Therefore what we have is a curve in three dimensions with parameter θ , where $0 \le \theta \le 2\pi$.

	Ω (in degrees)	i (in degrees)
Earth	0	0
Jupiter	99	1
Halley's comet	58	162

(a) On the same graph, using (4a, b, c), plot the orbits of the three objects listed in the preceding table. When doing this use the "view" option to restrict the range of the z-axis to $-0.3 \le z \le 0.3$. Identify the planets and comet on the graph (you will have to do this by hand).

(b) In this problem we will let $r_e(\theta)$ designate the position vector for the Earth and $r_h(\theta)$ the position vector for Halley's comet. When looking at the orbits in three dimensions, it is seen that Earth and Halley's comet can no longer collide. We want to determine, however, just how close they can get to each other. To do this you are to find the point in Earth's orbit (when $\theta = \theta_e$) and the point in the orbit of Halley's comet (when $\theta = \theta_h$) so that the distance

$$\| r_e(\theta_e) - r_h(\theta_h) \|$$

is as small as possible. Convince yourself that this task is the same as finding a point $(u, v) = (u_0, v_0)$ where the function

$$f(u, v) = \| r_e(u) - r_h(v) \|^2$$

takes on its minimum over $0 \le u \le 2\pi$ and $0 \le v \le 2\pi$. Once this point is found, then $\theta_e = u_0$ and $\theta_h = v_0$.

Find just how close these two objects can get to each other and do this in two steps:

(i) First, estimate where the minimum occurs. This can be done by plotting $f(u, v)$ over $0 \le u \le 2\pi$ and $0 \le v \le 2\pi$. From this plot establish approximate intervals for u and v that contain the minimum point (the intervals should have length 2 or less, and the critical point should be approximately in the center). This is going to require use of the "view" option for `plot3d,` and it is suggested that you plot ln(f) instead of f (this will make the min point a little easier to see). In your write-up explain why it is necessary to obtain this estimate for the intervals.

(ii) Use `fsolve` to find the point where the min occurs, and from this determine the minimum distance. By the way, for your information, the Moon is approximately 0.0025 AU from Earth (i.e., approximately one-quarter of a million miles).

(c) The closest Halley's comet got on its last appearance was about 0.4 AU (this occurred on April 11, 1986). Why do you think this is larger than your answer in (b)?

35

Understanding and Using Directional Derivatives

Objective

To use Maple for calculating directional derivatives and to illustrate them by examining a problem in heat conduction.

For a function of one variable $v = F(u)$, the derivative of F at a point $u = u_0$ is $F'(u_0)$, and this number is the rate of change of F at u_0. For a function of two variables $z = f(x, y)$, the graph is a surface. Does it make sense to ask, "what is the rate of change of f at a point $x = x_0$, $y = y_0$?"

To answer this question, imagine yourself standing on the graph of the surface at the point $x = x_0$, $y = y_0$, and $z = z_0 = f(x_0, y_0)$. This means you are at an "altitude" z_0 above the plane $z = 0$. If you start walking from this point, what is your rate of change of "altitude"? It might be the same no matter which way you start walking—if you are at the top of a rounded hill. Or, if you are part way up the hill, your rate of change may depend on which direction you start walking—for example, it will be positive if you go uphill and negative if you go downhill.

Therefore "the rate of change of $f(x, y)$ at a point" makes sense **only** if we specify a **direction** for the rate of change. This is one idea behind the concept of a **directional derivative**. Let the direction be given by a **unit** vector L, with two components (Lx, Ly) in the x and y directions. Also, let the vector G be the gradient of f evaluated at the point $P = (x_0, y_0)$. Then the directional derivative of f at P in the direction of L is given by the dot product of L and G. This number is the rate of change of f at P in the direction given by L.

In this Laboratory we use Maple to carry out the calculation of some directional derivatives, as well as to help visualize the solution surface that comes from a heat conduction problem. We will examine a function $T(x, y, t)$ that represents the value of temperature at a point (x, y) at time t. Note that T is a function of three variables. We will illustrate T by choosing particular values of t and then graphing T as a function of just the two variables x and y. Looking at several pictures allows us to imagine how the temperature changes with time in a portion of the x,y-plane, that is, how the "conduction of heat" takes place in that region.

Before we examine the heat flow problem, here is an example of using Maple to find directional derivatives.

EXAMPLE

Suppose $z = f(x, y)$ is the elliptical paraboloid given by

> f := 10 – (x – 1)^2 – (y + 1)^2/4;

We can plot the surface using `plot3d:`

> plot3d(f, x = –5..5, y = –10..10, axes = BOXED, title = `Paraboloidal Hill`);

Save this picture. Try to imagine being on the surface at the point $P = (-2, 4)$, for example. Can you visualize what the directional derivatives are at P?

To find directional derivatives at P, we first need the gradient there:

> with(linalg):
> g := grad(f, [x, y]);
> G := subs(x = –2, y = 4, op(g));

In the direction of the positive x-axis, L is the vector (1, 0), so the directional derivative is just the first component of G, i.e., f_x :

> DD1 := dotprod(G, vector([1, 0]));

From the graph, can you see why DD1 has a reasonably large positive value? In the direction of the line $y = x$, L is (1/sqrt(2), 1/sqrt(2)) :

> DD2 := evalf(dotprod(G, vector([1/sqrt(2), 1/sqrt(2)])));

Evaluate a few other directional derivatives if you like. In the direction of the positive y-axis, DD must be –2.5. So, somewhere between the 45-degree line and the positive y-axis, DD has become zero. Can you visualize this direction on the graph? Let us next determine it.

A unit vector in any given direction can be written as

> L := vector([cos(th), sin(th)]);

where th is the angle between the direction of interest and the x-axis (if you prefer, use theta instead of th). Therefore the directional derivative of f at the point (−2 , 4) in an **arbitrary** direction (specified by the angle th) is

> DD := dotprod(G, L);

Where is DD zero?

> solve(DD, th);

What is this angle in degrees?

> evalf(convert(", degrees));

Therefore at a little past 67 degrees, the rate of change of f at (−2 , 4) goes to zero, and it decreases as the direction approaches the positive y-axis.

As th continues to increase, the DD becomes more negative, until it reaches a minimum (negative) value. You can see this behavior from a plot of DD :

> plot(DD, th = −Pi..Pi, title = `Directional Derivative at P`);

From the plot, the minimum value is close to but less than π . What is this direction and the minimum rate of change?

> DDp := diff(DD, th);
> evalf(convert(fsolve(DDp, th, Pi/2..Pi), degrees));

This angle, of about 157 degrees, gives the smallest (negative) value of the DD . Can you visualize this direction on the surface?

> evalf(subs(th = fsolve(DDp, th, Pi/2..Pi), DD));

$$- 6.500000000$$

What is the meaning of this value? Recall that the components of G are (6 , −2.5) . The length of this vector is

> norm(G, 2);

This illustrates the fact that the minimum (negative) directional derivative is in the **opposite direction** from the gradient. In what direction does the maximum (positive) directional derivative occur? What is its value?

Look once again at the graph of the surface, so that you visualize the positive and negative behaviors of the directional derivatives at P.

Remember that for a smooth surface $z = f(x, y)$, the tangent plane at any point gives a good approximation to the surface. What is the tangent plane TP at the point P?

```
> TP := subs(x = -2, y = 4, f) + dotprod(G, vector( [x + 2, y - 4] ));
> plot3d({f, TP}, x = -5..5, y = -10..10, axes = BOXED,
>     title = `Hill and Tangent Plane`);
```

Here is another way to use Maple to get the tangent plane. Use the `mtaylor` command for a multivariable Taylor series:

```
> readlib(mtaylor);
> tanplane := mtaylor(f, {x = -2, y = 4}, 2);
```

Notice that the same equation for the tangent plane is constructed by the `mtaylor` command. Read more about it with ?mtaylor .

Since the function should be well approximated by its tangent plane at P, you should be able to visualize the positive and negative behaviors of the directional derivatives on the tangent plane. In fact, how are the DDs of the surface and the tangent plane related?

```
> DDTP := dotprod( grad(TP, [x, y]), L);
```

Do you understand why the directional derivatives of the surface at P and its tangent plane at P are the **same** ? Be sure you understand this before you start the problems.

Laboratory Problems

In Problem 1 the heat conduction problem is introduced. In Problem 2 directional derivatives are used to determine some of the characteristics of the solution.

1. Suppose that before the time $t = 0$ the x,y-plane is at a constant temperature, which for convenience we will call zero. However, at time $t = 0$, heat is added at the point $(0, 1)$—a "hot spot"—with temperature 2

at that instant. Also at $t = 0$, a "cold spot" is added at $(0, -1)$, with temperature -2 at that instant.

The theory of heat conduction states that at any later time $t > 0$ and for any point (x, y), the temperature $T(x, y, t)$ satisfies the following partial differential equation:

$$T_t = \kappa(T_{xx} + T_{yy}),\qquad\qquad\qquad (1)$$

where κ is called the "thermal conductivity." For this Laboratory we consider the case $\kappa = 1$. Think of time t as measured in minutes.

(a) Show that the following function T satisfies Eq. (1) by using Maple to compute its partial derivatives:

$$T = \frac{1}{t\sqrt{\pi}} \exp\left(- \frac{x^2 + (y-1)^2}{4t} \right) - \frac{1}{t\sqrt{\pi}} \exp\left(- \frac{x^2 + (y+1)^2}{4t} \right) .$$

(b) Plot the temperature for $t = 0.5$. Experiment to find good intervals of x and y to show the features of the surface. Then plot the temperature for four or five values of time t between 0.25 and 2. You may want to use a do-loop. Hand in your favorite plot, and indicate the value of t on the plot. Describe in your own words the behavior of the temperature with time.

(c) Find the location (i.e., the x- and y-coordinates) of the maximum value of the temperature for several values of t between 0.25 and 2. Explain the meaning of your results. Note that to do this problem you may want to use a do-loop. Also, when using `fsolve` restrict x to the interval $-1 < x < 1$ and restrict y to the interval $0 < y < 2.1$.

2. For all parts of this problem, use the single time value $t = 1$.

(a) Find the formula for the directional derivative of $T(x, y, 1)$ at any point (x, y) in an arbitrary direction specified by an angle th from the x-axis.

(b) Look at a few points on the y-axis, i.e., $x = 0$. At each of the points you pick, describe the behavior of the directional derivatives as a function of the angle th. Make sure your description is consistent with the 3D picture of the temperature surface.

(c) Consider the point $P = (1, 1)$. Make a plot of the temperature surface and the tangent plane at that point. Use the "view" option to restrict the vertical axis to between -1 and 1, and find an interesting orientation to view the surface and tangent plane. Hand in this plot.

(d) Find the directional derivative of T at the point P = (1 , 1).
Describe the behavior of the directional derivative as a function of the
angle th. Where is it a max, and where is it a min?

36

Discovering the Second-Order Optimality Conditions

Objective

To use second-degree Taylor polynomials to discover the second-order optimality conditions.

In this Lab we will investigate how second-degree Taylor polynomials can be used to approximate a function of two variables at a given point. When the given point is a stationary point (where the first partial derivatives f_x and f_y equal zero), we will see how this approximation can be used to discover the second-order optimality conditions that involve the second partial derivatives f_{xx}, f_{yy}, and f_{xy}.

A function $f(x, y)$ whose second partial derivatives are continuous can be approximated in the vicinity of a point (x_0, y_0) by the quadratic function

$$T(x, y) = f(x_0, y_0) + (x - x_0)f_x + (y - y_0)f_y$$

$$+ \frac{1}{2}[(x - x_0)^2 f_{xx} + 2(x - x_0)(y - y_0)f_{xy} + (y - y_0)^2 f_{yy}],$$

where each of the first and second partial derivatives is evaluated at the point (x_0, y_0). The function $T(x, y)$ is called the second-degree Taylor polynomial for f expanded about the point (x_0, y_0) . The term "second-degree" simply refers to the fact that the sum of the exponents in any term is at most two.

EXAMPLE

To illustrate how Maple can be used to find T , consider the function

$$f(x, y) = \frac{1}{9}y^3 + 3x^2y + 9x^2 + y^2 + xy + 9 .$$

To approximate f at the point $x_0 = -1$ and $y_0 = 1$, we can use the Maple command `mtaylor` to construct the function $T(x, y)$. First, it is necessary to read the command from the Maple library by entering

> readlib(mtaylor):

Of course, we need to enter f into Maple:

> f := y**3/9 + 3*x**2*y + 9*x**2 + y**2 + x*y + 9;

and then we can construct the second-degree Taylor polynomial T for f , at the point $(x_0 , y_0) = (-1 , 1)$, by entering

> T := mtaylor(f, [x = -1, y = 1], 3);

$$T := -56/9 + 13/3\ y - 23\ x - 5\ (x + 1)\ (y - 1) + 12\ (x + 1)^2 + 4/3\ (y - 1)^2$$

The 3 that appears in the `mtaylor` command tells Maple that we want the second-degree Taylor polynomial. One difference between the output of this command and what is obtained using `taylor` is that there is no remainder term. So, we will not have to use the `convert` command when using `mtaylor.`

The `mtaylor` command can also be used to construct a first-degree Taylor approximation by using a 2 in place of the 3 and entering

> w := mtaylor(f, [x = -1, y = 1], 2);

$$w := -56/9 + 13/3\ y - 23\ x$$

This first-degree Taylor approximation w is just the equation for the plane that is tangent to the graph of f at the point $(-1 , 1 , 190/9)$. Recall that the equation of the plane tangent to f at the point (x_0 , y_0) is given as

$$w = f(x_0, y_0) + (x - x_0)f_x + (y - y_0)f_y ,$$

where f_x and f_y are the first partial derivatives evaluated at (x_0 , y_0) . A discussion about using `mtaylor` to find the equation of a tangent plane can be found in the previous Lab.

Laboratory Problems

1. Plot the functions w and f on the same axes for $x = -1.5..1.5$ and $y = -7..1$ to check that w is indeed tangent to the graph of f at the point $(-1, 1)$. You do not need to hand in this plot. Now plot the functions w, f, and T on the same axes for $x = -1.5..1.5$ and $y = -7..1$. Notice that the second-degree Taylor polynomial T is a better approximation to f than w because T does a better job of accounting for the curvature of the graph of f.

    Discussion: We can find the stationary points by finding the first partial derivatives, f_x and f_y, and then solving the equations $f_x = 0$ and $f_y = 0$:

    > fx := diff(f, x); fy := diff(f, y);

    > solve({fx = 0,fy = 0}, {x, y});

    $$\{y = 0, x = 0\}, \{y = -6, x = -1/3\}, \{y = -7/2, x = -7/6\},$$
    $$\{y = -5/2, x = 5/6\}$$

2. Plot the function f over $x = -1.5..1.5$ and $y = -7..1$ and by looking at the graph classify each of the previous stationary points as a local maximizing point, a local minimizing point, or a saddle point.

    Discussion: You should have found that the point $(0, 0)$ is a local min point. Let's construct the second-degree Taylor polynomial for f expanded about $(0, 0)$ and see how good an approximation it is for f near this point.

    > T1 := mtaylor(f, [x = 0, y = 0], 3);

    $$T1 := 9 + xy + 9x^2 + y^2$$

    > plot3d({f, T1}, x = -1..1, y = -1..1, title = `Comparison with T1`);

    From this plot it appears that T1 is a satisfactory approximation to f over the region $x = -1..1$ and $y = -1..1$, but we cannot see how good the approximation is near the point $(0, 0)$.

3. Plot f and T1 over the smaller region $x = -0.1..0.1$ and $y = -0.1..0.1$ and notice that T1 approximates f very well over this region. You do not need to hand in this plot but do answer the following questions. Do you think that $(0, 0)$ is also a minimizing point for T1? Just answer yes or no. Show that $(0, 0)$ is the only stationary point for T1.

4. Use the `mtaylor` command to construct the equation for the plane tangent to f at $(0,0)$. Why is it that the plane is parallel to the x,y-plane? Also, use the `mtaylor` command to construct the equation for the plane tangent to T1 at the point $(0,0)$. Notice that the two equations are identical. On the same axes, over $x = -0.1..0.1$ and $y = -0.1..0.1$, plot f, T1, and the plane tangent to both at $(0,0)$.

Discussion: In general (x_0, y_0) is a stationary point for f if and only if it is a stationary point for the second-degree Taylor polynomial expanded about the point (x_0, y_0). To see that this is true, let $f(x_0, y_0) = f_0$ and recall from the previous definition that the second-degree Taylor polynomial expanded about (x_0, y_0) is defined as

> T := f0 + fx*(x − x0) + fy*(y − y0)

> + 0.5*(fxx*(x − x0)**2 + 2*fxy*(x − x0)*(y − y0) + fyy*(y − y0)**2);

where fx, fy, fxx, fxy, and fyy are evaluated at (x_0, y_0) and so are constants. Now take the partial derivatives of T :

> Tx := diff(T, x); Ty := diff(T, y);

$$Tx := fx + 1.0\,fxx\,(x - x0) + 1.0\,fxy\,(y - y0)$$
$$Ty := fy + 1.0\,fxy\,(x - x0) + 1.0\,fyy\,(y - y0)$$

Notice that when $x = x_0$ and $y = y_0$ then these two equations show that $T_x = f_x$ and $T_y = f_y$. Therefore $f_x = 0$ and $f_y = 0$ if and only if $T_x = 0$ and $T_y = 0$, and we can study the stationary point of f by studying the stationary point of T.

For example, to see why a stationary point (x_0, y_0) is a minimum point for the function f, we will first ask what is there about T that makes (x_0, y_0) a minimum point for T ? Look again at our definition of $T(x, y)$:

$$T(x, y) = f(x_0, y_0) + (x - x_0)f_x + (y - y_0)f_y$$

$$+ \frac{1}{2}[\,(x - x_0)^2 f_{xx} + 2(x - x_0)(y - y_0)f_{xy} + (y - y_0)^2 f_{yy}\,]\ .$$

The constant term $f(x_0, y_0)$ only raises or lowers the graph of T, so it is not important in determining whether (x_0, y_0) is a min point. If we used just the constant plus the linear terms,

$$w = f(x_0, y_0) + (x - x_0)f_x + (y - y_0)f_y\ ,$$

we would only get the plane $w = f(x_0, y_0)$, which is flat and parallel to the x,y-plane because $f_x = 0$ and $f_y = 0$ at the stationary point (x_0, y_0). It must be the second-order terms involving f_{xx}, f_{xy}, and f_{yy} that determine whether T rises up above this flat tangent plane or falls downward below the tangent plane.

The main goal of this Lab is to see whether we can discover what there is about the second partial derivatives f_{xx}, f_{xy}, and f_{yy} that makes a stationary point a min point, a max point, or a saddle point. To help in this investigation, there is a Maple command that computes a matrix that contains the second partial derivatives. This matrix is a 2 x 2 array with the form

$$ H = \begin{pmatrix} f_{xx} & f_{xy} \\ f_{yx} & f_{yy} \end{pmatrix}. $$

This matrix is called the Hessian matrix. For functions with continuous second partial derivatives it is always true that $f_{xy} = f_{yx}$, so this Hessian matrix is symmetric. We can easily compute the Hessian matrix using the `linalg` package in Maple. We get this package of routines by entering

> with(linalg):

If you enter this command and use a semicolon instead of a colon, you will see all the commands that are available. To calculate the Hessian matrix H for our current function f, we enter

> H := hessian(f, [x, y]);

To evaluate the Hessian matrix at a point, say at the saddle point $(5/6, -5/2)$, we enter

> H1 := subs(x = 5/6, y = -5/2, op(H));

Notice that we have to refer to the matrix H using op(H). To get decimals we use

> H1 := evalf(");

5. For each of the four stationary points that we found following Problem 1, find the Hessian matrix for our current function f. It will help for later questions if following each calculation of a Hessian matrix you indicate whether the graph shows the stationary point to be a min point, a max point, or a saddle point.

We will now look at some other functions and see what their Hessian matrices look like at their various stationary points.

6. Consider the function f(x , y) = 4*x**2*y – 2*x**4 – y**4 – 3 . Find the three stationary points for f . Ignore the complex roots. Plot f , and use the graph to classify each stationary point as a min point, a max point, or a saddle point. For each stationary point calculate the Hessian matrix, and for later use indicate whether it corresponds to a min, max, or saddle point.

7. Consider the function f(x , y) = x**4 + y**4 – 4*x*y + 5 . Find the three stationary points for f . Ignore the complex roots. Plot f , and use the graph to classify each stationary point as a min point, a max point, or a saddle point. For each stationary point calculate the Hessian matrix, and for later use indicate whether it corresponds to a min, max, or saddle point.

   Discussion: We are now ready to draw some conclusions from these problems. So far you should have collected the following list of Hessian matrices. You can use this list as a partial check on your previous work. Make sure you found all of these Hessian matrices for the three functions we have considered so far, namely:

$$f(x , y) = y**3/9 + 3*x**2*y + 9*x**2 + y**2+x*y + 9 ,$$
$$f(x , y) = 4*x**2*y – 2*x**4 – y**4 – 3 ,$$
$$f(x , y) = x**4 + y**4 – 4*x*y + 5 .$$

Hessian matrices that correspond to local minimizing points:

$$\begin{pmatrix} 12 & -4 \\ -4 & 12 \end{pmatrix}, \quad \begin{pmatrix} 18 & 1 \\ 1 & 2 \end{pmatrix}.$$

Hessian matrices that correspond to local maximizing points:

$$\begin{pmatrix} -18 & -1 \\ -1 & -18 \end{pmatrix}, \quad \begin{pmatrix} -16 & 8 \\ 1 & -12 \end{pmatrix}, \quad \begin{pmatrix} -16 & -8 \\ -8 & -12 \end{pmatrix}.$$

Hessian matrices that correspond to saddle points:

$$\begin{pmatrix} -3 & -6 \\ -6 & -1/3 \end{pmatrix}, \quad \begin{pmatrix} 3 & 6 \\ 6 & 1/3 \end{pmatrix}, \quad \begin{pmatrix} 0 & 0 \\ 0 & 0 \end{pmatrix}, \quad \begin{pmatrix} 0 & -4 \\ -4 & 0 \end{pmatrix}.$$

Remember, our goal is to see whether we can use the values of f_{xx}, f_{xy}, and f_{yy} in these Hessian matrices to classify a point as a local min point, a local max point, or a saddle point.

8. (a) Notice that $f_{xx} > 0$ for each of the local min points. If $f_{xx} > 0$ at a stationary point, does this always mean that the stationary point is a local min point? Answer true or false, and justify your answer based on the given list of Hessian matrices.

(b) Notice that $f_{xx} < 0$ for each of the local max points. If $f_{xx} < 0$ at a stationary point, does this always mean that the stationary point is a local max point? Answer true or false, and justify your answer based on the list of Hessian matrices.

(c) Calculate the determinants of each of the Hessian matrices. What property of the determinant of the Hessian seems to distinguish saddle points from local max or local min points?

(d) In general, it is true that the determinants of Hessian matrices for local min points and for local max points are positive. What property of the Hessian seems to distinguish local min points from local max points?

Discussion: Based on the preceding observations, two conclusions that appear to be true are the following:

If (x_0, y_0) is a stationary point for f, and if $f_{xx} > 0$ and det(hessian) > 0, then (x_0, y_0) is a local min point.

If (x_0, y_0) is a stationary point for f, and if $f_{xx} < 0$ and det(hessian) > 0, then (x_0, y_0) is a local max point.

These two conclusions are indeed true.

Another conclusion that appears to be true is the following:

If (x_0, y_0) is a stationary point for f, and if det(hessian) ≤ 0, then (x_0, y_0) is a saddle point.

This conclusion is not quite correct, as we will see from our final problem.

9. Consider the function $f(x, y) = x^{**}4 + y^{**}4$. The point $(0, 0)$ is clearly a min point. Show that its Hessian evaluated at $(0, 0)$ is given by

$$\begin{pmatrix} 0 & 0 \\ 0 & 0 \end{pmatrix}.$$

Also, show that the Hessian matrix for the function $f(x, y) = -x^{**}2 - y^{**}4$ evaluated at its max point $(0, 0)$ is given by

$$\begin{pmatrix} -2 & 0 \\ 0 & 0 \end{pmatrix}.$$

Thus, a local min point or a local max point can also have det(hessian) = 0. So at a stationary point, if det(hessian) = 0 we cannot make any conclusion about the point. It might be a local min, a local max, or a saddle point.

Discussion: We summarize this discussion by stating **the second-order optimality conditions**:

If (x_0, y_0) is a stationary point for f, and if $f_{xx} > 0$ and det(hessian) > 0, then (x_0, y_0) is a local min point.

If (x_0, y_0) is a stationary point for f, and if $f_{xx} < 0$ and det(hessian) > 0, then (x_0, y_0) is a local max point.

If (x_0, y_0) is a stationary point for f, and if det(hessian) < 0, then (x_0, y_0) is a saddle point.

If (x_0, y_0) is a stationary point for f, and if det(hessian) = 0, then no conclusion is possible without further analysis; (x_0, y_0) might be a local min, a local max, or a saddle point.

37

Double Integrals for Centers of Mass

Objective

To use double integrals for the calculation of the center of mass of a region in the x,y-plane.

One application of double integrals is the calculation of the center of mass of a region Ω in the x,y-plane. If the mass density in the region is $\rho(x\,,y)$, then the coordinates corresponding to the center of mass of that region are

$$\bar{x} = L_y/M\,,\qquad \bar{y} = L_x/M\,,$$

where the mass M and the first moments L_x, L_y about the x, y-axes are given by

$$M = \iint\limits_{\Omega} \rho\,dA\,,\quad L_x = \iint\limits_{\Omega} \rho y\,dA\,,\quad L_y = \iint\limits_{\Omega} \rho x\,dA\,.$$

In this Laboratory you will first work through an example in which the boundaries of the region are specified by four functions $y = yi(x)$, $i = 1,...,4$ and two segments of the x-axis. The density is a constant for these calculations. You will then carry out similar calculations for a nonuniform density and for differently sized regions. You will see the application of symmetry arguments to simplify the calculations.

Maple is essential both because the symbolic integral calculations are lengthy and because numerical values are not simple integers.

EXAMPLES

No new commands are introduced in this Laboratory. The iterative application of the `int` command is needed in order to calculate double integrals. The syntax is illustrated with commands in the following example.

The well-known "sign" of a certain retailer can be approximately modeled by the equations:

```
> y1 := 10 – 1.5*(x – 2)^2;          # for 0 <_ x <_ x1
> y2 := 9.75 – 2.25*(x – 2)^2;       # for 0 <_ x <_ x2
> y3 := 10 – 1.5*(x + 2)^2;          # for –x1 <_ x <_ 0
> y4 := 9.75 – 2.25*(x + 2)^2;       # for –x2 <_ x <_ 0
```

First let's plot these four curves, taking account of the ranges of the variables. We need the points of intersection of y1 and y2 with the ground (that is, the values of x when y = 0):

```
> x1 := fsolve(y1, x, 0..10);
> x2 := fsolve(y2, x, 0..10);
```

One way to plot this sign is as follows: [†]

```
> plot({[x, y1, x = 0..x1], [x, y2, x = 0..x2], [x, y3, x = –x1..0], [x, y4, x = –x2..0]},
>     x = –5..5, y = 0..10, title = `McDonald's Arches`);
```

The mass M of the sign can be expressed as a double integral. Assume first that the mass density of the sign is a constant, namely $\rho = 1$. The symmetry of the equations (and the plot) then show that M = 2*(mass of right side of sign) . You should **convince yourself** that:

```
> M := 2*(int( int(1, y = y2..y1), x = 0..x2)  +  int(int(1, y = 0..y1), x = x2..x1) );
```

With $\rho = 1$, this is also the area of the region occupied by the sign.

The coordinates (xbar , ybar) of the center of mass are given in terms of ratios of integrals. However, as in calculations of mass and area, it is often **very** useful to

---

[†] McDonald's is a registered trademark of the McDonald's Corporation.

take advantage of symmetry arguments because they can reduce the work considerably. In this example, convince yourself that xbar = 0 . We also know that

> ybar := Lx/M;

where Lx is the moment of the sign about the x-axis. We can express Lx as a double integral:

> Lx := 2*(int(int(y, y = y2..y1), x = 0..x2) + int(int(y, y = 0..y1), x = x2..x1));
> ybar;

$$4.936619221$$

Note that y1(0) = 4 . Therefore, the center of mass is located **off** the "physical" sign (above it, actually), for the case of uniform density.

Laboratory Problems

Problem 1 includes the calculation for the center of mass when there is a variable density. Problem 2 extends the calculations of the center of mass to signs of different sizes and shapes.

1. Suppose we want the center of mass of the sign to be lower, for reasons of stability. One way to accomplish this is to change the shape of the sign— but we might not want to change an internationally known symbol! Another way is to change the mass density of the construction material for the sign.

    Suppose the mass density rho is taken to be $1 + r*(1 − y/10)^2$, which includes an adjustable parameter r (r is assumed nonnegative) to account for different possible materials. Note that the double integrals for xbar and ybar must include rho in the integrand (as in the formulas at the start of the Lab).

    (a) First let's understand the "model" for the mass density. What is the behavior of rho with distance y above the ground? Find the maximum and minimum values of the density on the sign.

    (b) Now let's consider the center of mass for the sign with nonuniform mass density. As for the previous case of constant density, show that the coordinate xbar is zero.

    (c) Find a formula for ybar in terms of r .

(d) Plot ybar for $0 \leq r \leq 10$ and observe that the graph decreases. Confirm this behavior by computing a derivative. Find a value of r, and label it rcrit, so that for r > rcrit, the point (xbar, ybar) lies **on** the physical sign.

(e) Can the center of mass ever be located **below** the physical sign (and above the ground) for any values of r? Base your answer on your formula from part (c).

2. Now suppose the sign is to be made larger or smaller, and we permit its shape to be slightly modified (what you might call "Arches for the Twenty-First Century"). One way to do this is by using a scaling factor F in new equations for the right side of the sign:

> y1F := 10*F – 1.5*(x – 2*sqrt(F))^2;   # for $0 <\_ x <\_ x1F$

> y2F := 9.75*F – 2.25*(x – 2*sqrt(F))^2;   # for $0 <\_ x <\_ x2F$

Equations for the left side won't be needed here.

(a) Find x1F and x2F, which are the points where the sign touches the ground.

(b) Make a plot showing the right sides of **two** signs, one for the case when F = 1 and another for some choice of F > 1. Make a second plot showing the right sides of two signs, one for the case when F = 1 and another for some choice of F < 1. Then from your plots and the equations, describe the behavior of the sign's size and shape with varying values of F.

(c) Assuming the new sign density rhoF = 1, find the coordinates (xbarF, ybarF) of the center of mass. Interpret your answer.

(d) Now suppose the mass density of the construction material depends on y, so that   rhoF = 1 + r*(1 – y/(10*F))^2. In view of your answers to part (c) and to Problem 1, what are the formulas for (xbarF, ybarF) in terms of r?

38

The Order of Integration in Triple Integrals

Objective

To compute triple integrals for moments of inertia of a solid, using Maple for visualization and calculation.

Among the applications of triple integrals is their use in determining the moment of inertia of solid objects. For example, the moment of inertia for the final stage of a rocket is needed to predict its flight properties and stability during the placement of a satellite in orbit. If a solid S has mass density $\rho(x, y, z)$, then its moment of inertia I_z about the z-axis is given by

$$I_z = \iiint\limits_{S} (x^2 + y^2)\rho \; dV \; .$$

Similar expressions hold for other moments of inertia.

Triple integrals are evaluated using iterated integrals. The process is to write dV as a product of dx, dy, and dz, then set up suitable limits of integration, and finally perform the three "partial integrations." Maple can be a great help for the integration step, as you will see in this Laboratory. Before performing the calculations, however, the limits of integration are needed. You will also see that

Maple can assist in finding the limits, by allowing you to visualize more easily the regions of integration.

In writing a triple integral as a sequence of three iterated integrals, you must decide in what order to perform the integrations... first z , then x , finally y ; or first x , then z , finally y ; and so on, for a total of six possibilities. How to decide? The equations that describe the solid may help you decide which order is more convenient, and visualizing the solid may also help. It is important to realize what the theory of triple integrals reveals about this decision. For all reasonable density functions, the value of the moment of inertia is the **same** for **any** choice among the six orders of integration. We take advantage of Maple to illustrate this important result.

Before stating the Laboratory problem, we first consider an example of finding the moment of inertia of a simple solid. In the process, we set up and calculate I_z using several different orders of integration.

EXAMPLES

Suppose the solid S is a wedge that is bounded by the planes $z = 1 - y$, $z = 0$, $x = -1$, $x = 1$, $y = -1$, and $y = 1$. In this example we will assume $\rho = 1$. Although it is not too difficult to visualize this simple region without a computer plot, we construct a 3D graph of it with Maple in order to illustrate the procedure for more complicated regions.

> f := 1 − y;
> plot3d(f, x = −1..1, y = −1..1, axes = BOXED, title = `The Wedge`);

You are strongly urged to add axes to your 3D plots, either in the command line or through the on-screen menu. Save your plot window.

(a) First let's set up the integral for I_z using the integration order of z , then y, and finally x . To do this notice that the "projection" of S onto the x,y-plane is just the square defined by -1 ≤ x ≤ 1 , -1 ≤ y ≤ 1 . To help visualize this, recall that the projection onto the x,y-plane can be thought of as the "shadow" the solid makes if there is a light placed high above the solid (so the light rays are parallel to the z-axis). Convince yourself that if

> distsq := x^2 + y^2;

then I_z is given by

> Iz := int(int(int(distsq, z = 0..f), y = −1..1), x = −1..1);

(b) Next we set up the I_z integral with integration in the order y , x , z . This order **requires** projection onto the x,z-plane. Look again at your plot, and see that the shadow on the x,z-plane is also a square, this time $-1 \leq x \leq 1$, $0 \leq z \leq 2$. Convince yourself that I_z should be given by the expression

> Iz := int(int(int(distsq, y = -1..1- z), x = -1..1), z = 0..2);

(c) Finally, we set up the I_z integral with integration in the order x , z , y . This order requires projection onto the y,z-plane. Examine the plot and see that the shadow on the y,z-plane is a triangle. Convince yourself that I_z is

> Iz := int(int(int(distsq, x = -1..1), z = 0..1- y), y = -1..1);

(d) Set up the integral for I_z with integration in the order x , y , z .

If you execute any two or more of the triple integration commands for I_z , you will see that Maple produces the same value each time, just as promised.

Although this solid is relatively simple, and none of the integrals above are too complicated, we can summarize a few lessons about order of integration.

1. All the integrals have the same integrand, that is, the same function distsq*rho appears inside the triple integral. Changing the order does not affect the integrand.

2. Once the choice of the innermost (first) integration variable is decided, it is then necessary to find the shadow of the solid on the coordinate plane of the **other two** variables.

3. The limits of the second and third integration variables have to be determined by examining the shadow region. [There is a complication that can arise, as you will see in the Laboratory Problems.]

4. The outermost (last) integration limits must be constants. The middle limits can either be constants or functions of one variable (the last integration variable). The innermost (first) limits can either be constants or functions of one variable or functions of two variables (the middle and last integration variables).

Laboratory Problems

We imagine a simple model for the "nose stage" S of a rocket, using two functions to form the boundary of S . These two functions, given in terms of the independent variables x and z , are

$$y = 8 - z^2 - 2 x^2 \qquad \text{(a paraboloid)} , \qquad (1)$$

$$y = z^2 \qquad\qquad \text{(a parabolic cylinder)} . \qquad\qquad (2)$$

Problems 1 and 2 comprise two different ways to calculate I_z. As such they are independent of one another, although at the end of Problem 2 a statement is made that the result is the same as obtained in Problem 1.

1. (a) Since the boundaries of S are given in (1) and (2) in terms of functions of x and z, we will calculate I_z with the y integration done first. Plot the boundary surfaces given in (1) and (2). As in the example, be sure to include axes on your plot, and you may want to use the `plot3d` options "numpoints" and "view." Rotate the plot by changing the angle Phi [there are some interesting views for Phi = 90° and different values of Theta]. Print one graph that you believe best illustrates the situation, and label the surfaces by hand. Based on this graph, determine the innermost integration limits.

    (b) Determine the projection of S onto the x,z-plane, and find the other two integration limits. Use Maple to calculate the value of I_z.

2. (a) Calculate I_z with the z integration done first. To do this the first task is to visualize the solid with x and y as independent variables. Solve (1) and (2) for z, and call the (four) solutions z1 , z2 , z3 , and z4 . These four equations each specify portions of the boundaries of S. Plot them on a single 3-D picture. [The resolution may be a little ragged in places on the plot, but just imagine that the surfaces are smooth. An extra credit problem follows that provides a better graph of the surfaces.] Experiment with the 3-D picture to find reasonable plotting intervals for x and y. Rotate the picture by changing the angle Theta by 90 degrees or more. Print out one of your pictures, and label the surfaces.

    (b) The second task is to determine the innermost (z) limits of integration, with the help of your plots. There are two different sets of limits, depending on which of z1 , z2 , z3 , and z4 are the boundaries of S. [Since there are two different sets of z limits, you will need to write I_z as a sum of two integrals. This is the complication mentioned in the third lesson about the order of integration.] Determine the two equations for the two boundaries of the projection onto the x,y-plane. Also determine the equation for the curve in the x,y-plane along which the z limits of integration change. Produce a 2D plot of these three curves, and label them. Mark the z limits of integration on the appropriate regions of the plot.

    (c) The final task is to set up the integrals for I_z including the remaining two limits of integration. Use whichever order x , y or y , x that you prefer. Then use Maple to evaluate these integrals, and thereby show that the same answer is obtained as in Problem 1(b).

EXTRA CREDIT PROBLEMS

1. The 3D plot in Problem 2(a) can be improved by representing the surfaces parametrically. Here is one set of parametric equations:

   > s1:= [sqrt(8 − t)*cos(s)/sqrt(2), t, sqrt(8 − t)*sin(s)];

   > s2:= [−sqrt(8 − t)*cos(s)/sqrt(2), t, −sqrt(8 − t)*sin(s)];

   > s3:= [−2*s/Pi, t, sqrt(t)];

   > s4:= [−2*s/Pi, t, −sqrt(t)];

   A command which produces a plot using these equations is:

   > plot3d({s1, s2, s3, s4}, s = −Pi..Pi, t = 0..8);

   (a) Construct the plot. In what way is it an improvement over your plot from Problem 2(c)? How do the parametric equations make an improvement?

   (b) Explain carefully which of {s1 , s2 , s3 , s4} correspond to each of {z1 , z2 , z3 , z4}. Justify your answers. Don't forget to include the limits of the parameters s and t in your explanation.

2. Set up and evaluate integral(s) for I_z with the x integration done first. Show that the same result is obtained as in Problems 1(b) and 2(c).

39

An Application of Triple Integrals

Objective
To calculate the center of mass, using cylindrical coordinates, for different types of ice cream cones.

An ice cream cone on a hot summer day is a treat—so long as the ice cream stays in the cone, or on top of it. Did you ever suffer the trauma of having ice cream fall off the cone? If so, you might wonder what insight (if any) calculus offers about this dilemma.

One sure way to lose the ice cream is to tilt the cone at a large angle from the vertical. How large an angle? Although this question is clearly of importance to cone eaters, we will start with a simpler one. Suppose that the ice cream is fairly solid, and that you hold the cone nearly vertical (at least, much straighter than a three-year old would). Will the ice cream stay in place? You might think it should, even when you do not grip the cone tightly, provided that the center of mass of the ice cream is below the level of the cone. Such a "well-engineered" ice cream cone would presumably be stable to the inevitable small deviations from the vertical. So we are led to ask, where is the center of mass of (a mathematical model for) an ice cream cone?

The coordinates $(\overline{x}, \overline{y}, \overline{z})$ of the center of mass of a solid S are defined in terms of triple integrals. For example, if the solid has mass density $\rho(x, y, z)$, then $\overline{z}$ is

$$\overline{z} = \frac{\iiint_S z\rho \, dV}{\iiint_S \rho \, dV} \quad ,$$

and similarly for $\overline{x}$ and $\overline{y}$. We ignore density variations for ice cream, so $\rho = 1$ has no influence on the center of mass (put another way, we want to calculate the centroid of the ice cream cone).

Another simplification we make is that the ice cream cone has an "ideal" symmetric shape, so that it has an axis of symmetry. If we assume that this line of symmetry is the z-axis, then it follows that the center of mass lies on it, that is, both coordinates $\overline{x}$ and $\overline{y}$ are zero. Only $\overline{z}$ remains to be found. Furthermore, the line of symmetry means that cylindrical coordinates (r, θ, z) are likely to be useful in the calculations. Before tackling the ice cream cone, we illustrate the use of cylindrical coordinates in Maple with a simpler example.

EXAMPLE

Determine the center of mass of a truncated cone, with height h and base radius a. Assume the line of symmetry of the cone is the z-axis, and the apex is at the origin. Therefore the boundary surface is given by $z = hr/a$. We are keeping $\rho = 1$.

We can use Maple to produce a plot of the cone. With cylindrical coordinates, it is convenient to use the `cylinderplot` command. To use this, first read in the `plots` package.

```
> with(plots);
```

Remember that for new commands, "help is just a ? away."

```
> ?cylinderplot;
```

The examples from the help page show us that cylindrical plots use functions expressed as $r = f(\theta, z)$. So for the cone,

```
> F := a*z/h;
```

To plot the special case $h = 3$ and $a = 1$, use the command:

```
> cylinderplot(subs(a = 1, h = 3, F), theta = 0..2*Pi, z = 0..3, axes = BOXED,
>    title= `Cone`);
```

To compute the center of mass, we set up two triple integrals. We will integrate over z first, so we solve the cone equation for z.

> G := solve(F − r, z);

$$G := \frac{r\,h}{a}$$

Remember to include the factor r (from the cylindrical volume element) in the integrand. From the plot the limits of integration in z should be z = G to z = h, so the mass (or, what is the same thing here, the volume) is

> vol := int(int(int(r, z = G..h), r = 0..a), theta = 0..2*Pi);

$$vol := 1/3 \; a^2 \; h \; Pi$$

Does this formula look familiar? Next we need the moment about the plane z = 0,

> mom := int(int(int(r*z, z = G..h), r = 0..a), theta = 0..2*Pi);

$$mom := 1/4 \; a^2 \; h^2 \; Pi$$

> zbar := mom/vol;

$$zbar := 3/4 \; h$$

Notice that this result means (1) the center of mass of a circular cone is 3/4 the way up from the apex, and (2) the center of mass does **not** depend on the radius of the base of the cone—it only depends on the height h of the cone.

Laboratory Problems

In these problems you are to find $\bar{z}$ for various types of ice cream cones. In all cases the cone to be used is the one examined in the previous example.

In Problems 1 and 2 two different single-dip cones are considered. Consequently these problems are largely independent of one another, although the method for plotting an ice cream cone is described in Problem 1(b). It is not necessary to do Problem 1(b) in order to understand the explanation. In Problem 3 a double-dip ice cream cone is examined, and in this case the results from both Problems 1 and 2 are used.

1. Let's find the center of mass of a "canonical" ice cream cone. This is the model with a hemisphere of ice cream on top of the inverted cone. Also, we assume that the cone portion is **completely** filled with ice cream.

(a) Write down the equation for the upper half of a sphere with center at $r = 0$, $z = h$, and radius a. Write that equation in two forms, $z = F1(r)$ and $r = G1(z)$.

(b) Use Maple to display the canonical ice cream cone. You need to plot both G and G1 on the same picture using cylinderplot (the G here is the same one discussed in the example). You should **not** use a single cylinderplot command, because the ranges of values for the variable z in G and in G1 are different. So, what you do is generate plot files for the two surfaces individually. Here is one of the commands, for the example of $a = 1$ and $h = 3$:

```
> cone := cylinderplot(subs( a = 1, h = 3, G), theta = 0..2*Pi, z = 0..3,
>    axes = BOXED):
```

Make sure to use the colon at the end to suppress the output of number pairs. Construct a similar plot file for the ice cream. Finally, use the command `display3d` to show both surfaces together. Move the angle Phi to near 90° to show the canonical cone.

(c) Use functions F and F1 to calculate the volume and the moment of the canonical cone. Then find its center of mass.

(d) The center of mass formula is a function of a and h. Substitute $a = $ alpha*h into your formula and simplify it. Here alpha is the ratio of maximum width a to cone height h for the canonical cone. Now use the simplified formula to answer the burning question: For what values of alpha is the center of mass **inside** the cone? [You might find a plot helpful.]

(e) What are reasonable values of alpha for canonical ice cream cones? If you need assistance, ask for help at a friendly ice cream parlor. Is the center of mass always inside the cone? Don't forget to check big "waffle" cones.

2. It takes skill to make a canonical ice cream cone—to get the mound of ice cream to be nearly a hemisphere. How about a model for a cone that is not so symmetrical? One possibility is to assume that the ice cream mound is better approximated by half an ellipsoid.

(a) Write down the equation for the appropriate ellipsoid that has two semiaxes a and b, and center at $z = h$, $r = 0$. Note the total height of the ice cream cone is now $h + b$. Write the equation in two forms, $z = F2(r)$ and $r = G2(z)$.

(b) Use Maple to display the ice cream cone for $a = 1$, $b = 2$, and $h = 3$. The method that should be used to do this is described in Problem 1(b).

(c) Find the center of mass.

(d) Simplify your center of mass formula. How does your formula depend on a? On b? For what values of b is the center of mass **inside** the cone? Do you need any more field work at the ice cream parlor?

3. An all-time favorite ice cream cone is the double-dipper, with a second scoop on top of the cone. However, this kind is more susceptible to disaster than the cones in Problems 1 and 2. One model for a double-dipper is to use the ellipsoidal cone from Problem 2 and add a sphere of radius a on top. There are now other considerations besides the location of the center of mass: what if the second scoop slips off the top, and so forth. But we still hope to have the center of mass inside the cone, so let's compute its location.

(a) We could use the same method as in Problems 1 and 2 to compute the center of mass. However, here is a simpler way. Suppose the mass and z-coordinate of the center of mass for the ellipsoidal cone are Me and zbare, and Ms and zbars are the same numbers for the sphere. Then the center of mass for the double-dipper, z = zbardd, is given by

$$\text{zbardd} = (\ (\text{zbare})(\text{Me}) + (\text{zbars})(\text{Ms})\ )/(\text{Me} + \text{Ms})\ . \qquad (1)$$

Use triple integrals (not Maple) to show why Eq. (1) is true.

(b) Use (1) to compute zbardd, using zbare and Me from Problem 2. Simplify your formula as much as possible. Your zbardd will be a function of h, b, and a.

(c) The requirement that the center of mass is **inside** the cone is an inequality. Set up this inequality, and simplify it as much as possible. If you make the substitution b = (beta)*a, you should be able to express the requirement in the form h > a ∗ f(beta), where f(beta) is some function of beta that you should find. [Remember that Maple has more skills in manipulating equalities than inequalities, so using Maple it is better to analyze the case when zbardd is just at the height h of the cone and then to express the inequality condition.]

(d) Explain the significance of your final result from part (c). What does the function f(beta) look like? What are reasonable values of h, a, and b for double-dippers? Is the center of mass always inside the cone for the reasonable values? Is more field work needed?

Appendix 1

Regrouping a Conditionally Convergent Series—Does Scalar Addition Always Commute?[†]

Given a series of numbers, it is natural to assume that you can add the numbers together any way you like and still obtain the same result. That is to say, we've always learned that scalar addition commutes. Well this is certainly true if the series is finite. For example, we know that $a + b + c = a + c + b = b + a + c$, etc. In fact, this statement also holds for the terms of an absolutely convergent infinite series. However, it is not necessarily true for a conditionally convergent series. This was first proved in a theorem due to Riemann, which loosely states that you can rearrange the terms in any conditionally convergent series and make it converge to any number you want!

Before turning to Maple, let's first try to determine why this is so and hopefully demystify this rather counter intuitive property of a conditionally convergent series. We begin by considering the example of the alternating harmonic series,

---

† Thanks to Edward Damiano (RPI) for the procedure `addto` and his contributions to this appendix.

$$\sum_{k=1}^{\infty} \frac{(-1)^{k+1}}{k} \ .$$

This converges, with order preserved, to ln(2). To illustrate how we are going to reorder this series note that

$$\sum_{k=1}^{\infty} \frac{(-1)^{k+1}}{k} \ = 1 - \frac{1}{2} + \frac{1}{3} - \frac{1}{4} + \frac{1}{5} - \frac{1}{6} + \dots \ = \ln(2)$$

and

$$\frac{1}{2} \sum_{k=1}^{\infty} \frac{(-1)^{k+1}}{k} \ = \frac{1}{2} - \frac{1}{4} + \frac{1}{6} - \frac{1}{8} + \frac{1}{10} - \dots$$

$$= 0 + \frac{1}{2} + 0 - \frac{1}{4} + 0 + \frac{1}{6} + 0 - \frac{1}{8} + 0 + \frac{1}{10} - \dots$$

$$= \frac{1}{2} \ln(2) \ .$$

These two series are now going to be added together term-by-term, and so

$$\sum_{k=1}^{\infty} \frac{(-1)^{k+1}}{k} \ + \ \frac{1}{2} \sum_{k=1}^{\infty} \frac{(-1)^{k+1}}{k}$$

$$= 1 + 0 - \frac{1}{2} + \frac{1}{2} + \frac{1}{3} + 0 - \frac{1}{4} - \frac{1}{4} + \frac{1}{5} + 0 - \frac{1}{6} + \frac{1}{6} + \frac{1}{7} + \dots$$

$$= 1 + \frac{1}{3} - \frac{1}{2} + \frac{1}{5} + \frac{1}{7} - \frac{1}{4} + \frac{1}{9} + \dots \ .$$

What is significant here is that this series adds up to $(3/2) \ln(2)$, and yet it contains exactly the same terms as the alternating harmonic series.

```
# The procedure `addto(x)` rearranges the terms in the alternating harmonic
# series (which converges to ln(2) when the order of the terms is preserved)
# so that the new series converges to  x .
addto:=proc(x)
    local goal, approxsn, m1, m2, n1, n2, i, j, k, Sn, terms, n, partsums,indx,jndx:
    goal:=evalf(x):
    approxsn:=0:  a:=`a`:  indx(0):=1:  jndx(0):=1:
```

```
m1:=1: m2:=1: n1:=1: n2:=1:
for i from 1 to 20 do
     while(approxsn <= goal) do
     a(i) := sum(1/(2*k-1), k=m1..n1);
     approxsn := evalf(sum(a(j), j=1..i));
               indx(i) := 2*n1-1;
               jndx(i) := 2*m1-1;
     n1:=n1 + 1;
     od;
     m1:=n1: i:=i+1:
     while(approxsn >= goal) do
     a(i) := - sum(1/(2*k), k=m2..n2);
     approxsn := evalf(sum(a(j), j=1..i));
               indx(i) := 2*n2;
               jndx(i) := 2*m2;
     n2:=n2 + 1;
     od;
     m2:=n2:
od:
Digits:=5:
terms := evalf([ [jndx(k), indx(k), a(k)] $k=1..20]):
Sn := n -> sum(a(k), k=1..n):
partsums := evalf( [ [n, Sn(n)] $n=1..20] ):
lprint(`The new order is:`);
lprint(`[first index, last index, sum of every other term over this range]`);
lprint();
print(terms);
lprint();
lprint(`The partial sums are:`);
lprint(`[new index, current partial sum]`);
lprint();
print(partsums);
plot({partsums, goal}, 0..20, style=LINE);
end:
```

EXAMPLES

It is assumed the commands for the procedure `addto` are contained in the file reorder .

```
> read reorder;
> addto(ln(2));   # ln(2) = 0.693147...
```

The new order is:

[first index, last index, sum of every other term over this range]

[[1., 1., 1.], [2., 2., -.50000], [3., 3., .33333], [4., 4., -.25000],

   [5., 5., .20000], [6., 6., -.16667], [7., 7., .14286], [8., 8., -.12500], ...

The partial sums are:
[new index, current partial sum]

  [[1., 1.], [2., .50000], [3., .83333], [4., .58333], [5., .78333],

    [6., .61667], [7., .75952], [8., .63452], [9., .74563], [10., .64563], ...

># Notice the order of the terms in the preceding series is preserved
># since ln(2) is the limit of the alternating harmonic series.

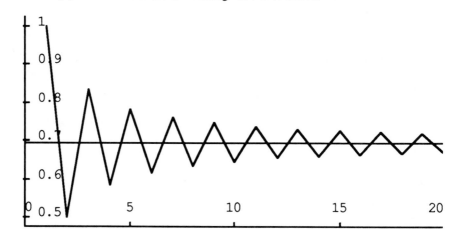

> addto((3/2) * ln(2));   # 3/2 ln(2) = 1.039720...

The new order is:
[first index, last index, sum of every other term over this range]

  [[1., 3., 1.3333], [2., 2., -.50000], [5., 7., .34286], [4., 4., -.25000],

    [9., 11., .20202], [6., 6., -.16667], [13., 15., .14359], ...

The partial sums are:
[new index, current partial sum]

[[1., 1.3333], [2., .83333], [3., 1.1762], [4., .92619], [5., 1.1282],

[6., .96154], [7., 1.1051], [8., .98013], [9., 1.0916], [10., .99159], ...

># Notice the order of the terms in this series as compared
># with the result obtained earlier. As before, we take two terms
># from the odd series for every one term we take from the even series.

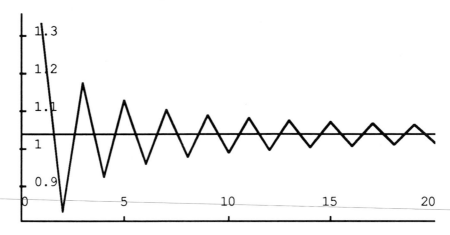

Appendix 2

Printing and Saving 2D Plots[†]

Printplot

The following procedure, `printplot,` will print a 2D plot on a laser printer. It was written for a UNIX workstation running X-windows, and so it may have to be modified to run on other systems.

```
#
#     Procedure to send a Maple plot to a temporary file, print
#     the file on the (default) laser printer and then delete file.
#
printplot:=proc()
 local unixcmd:
 if (nargs < 1) then
     print(` *** PRINTPLOT requires a valid plot request.`):
```

[†] Thanks to Mark Miller (RPI), and Geoff Aldis and Glenn Fulford (Australian Defense Force Academy) for their contributions to these procedures.

```
        print(`    Example: printplot(cos(x),x=0..Pi);`):
        ERROR(``):
    else
        if (traperror(plot(args[1..nargs])) = lasterror) then
            print(` *** PRINTPLOT found an invalid plot request.`):
            print(`    Example: printplot(cos(x), x=0..Pi);`):
            ERROR(``):
        else
            interface(plotdevice=postscript, plotoutput=`tmp8888818.ps`):
            print(plot(args[1..nargs])):
            interface(plotdevice=x11, plotoutput=terminal):
            unixcmd := `lpr -r tmp8888818.ps`:
            system(unixcmd):
        fi:
    fi:
    ``:
end:
```

EXAMPLES

```
> f := sin(x);
> printplot(f, x = 0..Pi, title = `An Example`);
> printplot({[t, t^2, t = 0..1], [t^3, 1 + t, t = -1..1]}, title = `Another Example`);
```

Saveplot

The following procedure, `saveplot,` will save a 2D plot in a specified file. It was written for a UNIX workstation running X-windows, and so it may have to be modified to run on other systems.

```
#
#    Procedure to save a Maple plot in a specified file.
#    No hardcopy printout.
#
saveplot:=proc()
    local filename:
    if (nargs < 2) then
        print(` *** SAVEPLOT requires a filename and a plot request.`):
        print(`    Example: saveplot(`maple.ps``, cos(x), x=0..Pi);`):
        ERROR(``):
    else
```

```
         if (traperror(plot(args[2..nargs])) = lasterror) then
                 print(` *** SAVEPLOT found an invalid plot request.`):
                 print(`    Example: saveplot(`maple.ps`, cos(x), x=0..Pi);`):
                 ERROR(` `):
         else
                 interface(plotdevice=postscript, plotoutput=args[1]):
                 print(plot(args[2..nargs])):
                 interface(plotdevice=x11, plotoutput=terminal):
         fi:
     fi:
end:
```

EXAMPLES

```
> f := sin(x);
> saveplot( `plotfile`, f, x = 0..Pi, title = `An Example`);
> saveplot( `plot2`, {[t, t^2, t = 0..1], [t^3, 1 + t, t = -1..1]}, title = `Another Example`);
```

Appendix 3

Plotting Volumes of Revolution

The following procedure, `plotR,` will make 3D plots of surfaces of revolution.

```
interface(echo = 4):
#
#     The command  plotR();  makes a 3D plot of the surface of revolution that
#     is obtained by rotating  y = f  around the x-axis, and also rotating  y = g
#     around the x-axis, for  a < x < b .
#
#     Before issuing this command  a , b , f , and  g  must be specified.
#     If you only want to rotate a single curve let  g := 0 .
#
interface(echo = 0):

plotR := proc()
local t,y,z,yy,zz,yyy,zzz:
yy := f*cos(t):
zz := f*sin(t):
yyy := g*cos(t):
zzz := g*sin(t):
if g=0 then
      plot3d([x, yy, zz], x = a..b, t = 0..2*Pi, labels = [x,y,z])
else
```

```
        plot3d({ [x, yy, zz], [x, yyy, zzz] }, x = a..b, t = 0..2*Pi, labels = [x,y,z])
fi:
end:
```

EXAMPLES

It is assumed the last procedure is contained in the file plotR .

```
> read plotR;
> a:=0:  b:=1:   f:=sqrt(x):  g:=x^2:
> plotR();
> a:=0:  b:=1:   f:=x^3: g:=0:
> plotR();
```